Order from Chaos

Theoretical Principles and Practical Aspects of the New Class of High-Entropy Materials

Luca Spiridigliozzi

Department of Civil and Mechanical Engineering – DICEM
University of Cassino and Southern Lazio (UNICAS)
Cassino (FR), Italy

CRC Press is an imprint of the
Taylor & Francis Group, an **informa** business

Cover credit: Image created by Luca Spiridigliozzi by modifying the output of DALL-E from Chat GPT Artificial Intelligence.

First edition published 2025
by CRC Press
2385 NW Executive Center Drive, Suite 320, Boca Raton FL 33431

and by CRC Press
4 Park Square, Milton Park, Abingdon, Oxon, OX14 4RN

Library of Congress Cataloging-in-Publication Data (applied for)

ISBN: 978-1-032-38794-9 (hbk)
ISBN: 978-1-032-38798-7 (pbk)
ISBN: 978-1-003-34680-7 (ebk)

DOI: 10.1201/9781003346807

Typeset in Times New Roman
by Prime Publishing Services

Acknowledgements

- I dedicate this work first and foremost to my little daughter, to make her proud of her dad when she will grow up.
- To my wife, the woman I have chosen to have by my side and with whom I want to continue to share the joys of my life.
- To my parents, who have shaped me to the man I am now with immeasurable love and trust.
- To all the people that over the course of these years has masterfully guided me along the winding path of scientific research, allowing me to meet many fantastic people who have been a part of it.
- To all my colleagues and friends, with whom I have shared many moments of everyday life and who gave me the strength to persevere in my work.

Preface

Entropy is a multifaceted concept that weaves into the fabric of our universe, encompassing everything from the intricate dance of molecules to the encoding of digital information in our modern world. This book seeks to explore the myriad applications of entropy, culminating in a focused examination of the exciting and emergent field of high-entropy oxides.

In thermodynamics, entropy embodies disorder and randomness, guiding the transformation of energy and the course of natural processes. Information theory coopts entropy as a measure of uncertainty, vital in modern communication systems and data handling. The world of materials science and chemistry utilizes entropy to understand phase behaviors, alloy formation, and chemical reactions. Engineers leverage entropy in various applications, including structural systems and energy efficiency.

Thus, in a generalized sense, entropy can be seen as a measure of uncertainty, randomness, or complexity in various systems and contexts. It plays a fundamental role in our understanding of nature and is central to many scientific and engineering disciplines.

Yet, at the forefront of these diverse applications lies a novel and burgeoning area of research: high-entropy oxides. This class of materials, characterized by their complex and disordered structure, has challenged traditional understanding and opened up new horizons in materials science. Their unique properties, derived from the high entropy of their compositions, offer promising potential for applications ranging from energy storage to catalysis and beyond.

This book is more than a journey through the complexities and beauty of entropy; it's a focused exploration into high-entropy oxides, where tradition meets innovation. Readers will be introduced to the fundamentals of entropy before diving into the cutting-edge research and technological advances surrounding these fascinating materials.

Whether you are a student, a seasoned scientist, or simply a curious mind, this book provides an insightful and comprehensive look into both the well-established principles of entropy and the thrilling frontier of high-entropy oxides. From the fundamental principles to the latest research findings, this volume serves as a guiding light, illuminating the path towards understanding and harnessing the tremendous potential of these remarkable materials.

Luca Spiridigliozzi

Contents

SECTION 1

The Concept of Entropy

Introduction

Entropy in Physics and Thermodynamics—Bridging Microscopic and Macroscopic Worlds

Entropy is a central concept in various scientific domains, but its essence lies in quantifying the degree of disorder, randomness, or unpredictability within a certain system.

In the realm of thermodynamics, entropy describes the measure of a system's microscopic states, often called configurations, which correspond to a certain macroscopic state. The macroscopic state of a system is characterized by observable properties, such as temperature, pressure, and volume. In contrast, the microscopic state pertains to the specific positioning and movement patterns of individual particles in that system.

Mathematically, entropy (S) of a thermodynamic system is given by:

$$S = k_B \cdot \ln(W) \tag{1}$$

where k_B denotes the Boltzmann constant, a fundamental constant of nature linking an individual particle's energy to the system's temperature, while W stands for the number of microscopic configurations that align with a specified macroscopic state. The natural logarithm in Equation (1) captures the sheer vastness of potential configurations even for modestly-sized systems.

The Second Law of Thermodynamics [1] posits that the entropy of an isolated or closed system will never decrease; it will either increase or remain unchanged. This relentless drive towards increasing entropy manifests in observable phenomena, such as heat always flowing spontaneously from regions of higher temperature to those of lower temperature. A system with maximal entropy, having reached a state of thermal equilibrium, no longer possesses the potential to perform useful work. Essentially, all parts of the system are in balance, with no gradients or differences to drive processes.

Entropy, however, is not just a mere parameter, as it provides deep insights into the nature of energy transfer and transformation. In a more philosophical lens, it offers a perspective on the inherent tendency of natural processes to evolve towards a state of maximum randomness or disorder, echoing the inexorable *"march of time"*.

Moreover, while entropy in thermodynamics connects with heat, energy, and work [2], the same concept, when transposed to information theory, measures the unpredictability or uncertainty of information. This flexibility of the entropy concept across domains underscores its foundational importance in science. In practical terms, understanding the thermodynamic concept of entropy has direct implications in various sectors, from designing efficient engines to predicting the behavior of materials under different conditions. Grasping the nuances of entropy is pivotal for advancements in areas like renewable energy, where managing and understanding energy flows is paramount.

Definitely, entropy serves as a bridge, connecting the microscopic intricacies of a system to its macroscopic behavior, and its understanding is fundamental to harnessing the power of thermodynamic processes effectively.

Entropy in Chemistry—Shaping Equilibria and Chemical Reactions

Entropy, a foundational concept in thermodynamics, permeates deeply into the heart of chemistry too. It captures the essence of disorder, chaos, or randomness that a system exhibits at the microscopic level. At its core, entropy provides insights into the myriad configurations or arrangements particles can assume while keeping a system's overarching properties—like temperature and pressure—consistent.

In the realm of chemical reactions, entropy changes provide clues about the reaction's nature and feasibility [3]. The following equation:

$$\Delta S = \Sigma S \,(\text{products}) - \Sigma S \,(\text{reactants}) \tag{2}$$

can be used to calculate the entropy change for chemical reactions. Here, ΣS denotes the aggregate entropy of either the reactants or products.

The direction and magnitude of entropy changes can dictate the spontaneity of reactions. Generally, reactions that usher in an increase in system entropy are spontaneous. This arises from nature's innate leaning towards states of higher randomness or disorder.

In chemical contexts, entropy also plays a pivotal role in shaping the understanding of reaction kinetics, equilibrium, and thermodynamics. Furthermore, it is instrumental in elucidating why certain reactions proceed while others do not, and under what conditions they might occur optimally.

Entropy in Information Theory—Decoding Uncertainty

In the vast landscape of information theory, entropy helps quantifying the randomness or uncertainty imbued in a dataset or a message [4]. Intrinsically, it gauges the richness of information or, in other words, the insights gleaned from observing or analyzing a particular dataset or message.

For a discrete random variable X governed by a probability mass function $P(X)$, entropy (H) is represented as:

$$H(X) = -\Sigma P(X) \times \log 2 P(X) \tag{3}$$

where Σ comprises the summation over all feasible values of X.

The base 2 logarithm, i.e. log 2, is used to present entropy in the unit of "bits."

Through Equation (3), it is possible to discern the mean quantum of information,which each symbol within the dataset or message conveys. Elevated entropy signifies heightened uncertainty or randomness in the data.

When every symbol in a dataset manifests with equivalent probability, entropy reaches its maximum. The maximal entropy for a set with n distinct symbols is:

$$H_{max} = \log 2 (n) \tag{4}$$

A dataset with such maximal entropy exudes complete randomness, where no symbol bears a greater likelihood than another.

In information theory, entropy has several practical implications, among which are the following:

- *Data Compression:* Entropy serves as a metric to gauge the compressibility of data. Data replete with redundancy or predictability can be compacted substantially without compromising on the embedded information. Thus, entropy offers a yardstick against which the adeptness of data compression algorithms can be evaluated.
- *Cryptography:* Within the domain of cryptography, entropy is pivotal. A robust cryptographic protocol should exhibit high entropy, ensuring that deciphering the encrypted message becomes an arduous task, even when harnessing considerable computational might.

Entropy in Crystallography—Crystals as Messages to Decipher

Entropy in crystallography is fundamentally a reflection of the degree of disorder within a crystalline structure (in a perfectly structured crystal, every atom or molecule is neatly and predictably organized) [5]. The more this ideal order is disrupted, either through structural imperfections, introduction of impurities, or increased temperature, the higher the entropy becomes. Atoms and molecules now have much more ways they can be arranged, and the system's predictability diminishes.

As already stated for the information theory, entropy quantifies the uncertainty or randomness of a data set or message. A message that is highly predictable (like a string of repeating characters) has low entropy. In contrast, a message with a diverse and unpredictable assortment of characters has high entropy, demanding a more intricate scheme to describe or encode it without loss. But considering the crystal as a message, where every atom or molecule represents a symbol or character, a perfectly structured crystal corresponds to a highly repetitive and predictable message, carrying low entropy. Conversely, introducing imperfections or disorder, the 'message' of the crystal becomes more complex and diverse, increasing its entropy.

Furthermore, just as in information theory where the aim is to encode messages in the most efficient way (minimizing redundancy), in crystallography, nature tends to favor structures that are energetically favorable. Energetically favorable structures tend to be more ordered (lower entropy) while higher energy states tend to be more chaotic (higher entropy).

To extend the analogy, the Random Close Packing (RCP) theory [6] in crystallography, which describes the optimal packing of spheres in a given volume, mirrors the quest in information theory to transmit the most data with the fewest bits. Thinking of each sphere or particle as a piece of information, the RCP theory is effectively trying to determine the most efficient way to 'pack' that information. When spheres pack together in an RCP arrangement, there is no long-range order, echoing a higher entropy state. This can be juxtaposed against the more ordered arrangements in crystal structures, which exhibit predictable patterns and sequences, reminiscent of lower entropy messages in information theory.

Definitely, both in crystallography and information theory, entropy emerges as a measure of unpredictability or randomness. While the mediums differ (crystals *vs* messages), the underlying principle remains consistent: systems tend towards states of lower energy and higher probability, balancing the intricate dance between order and chaos.

By understanding the concept of entropy in both realms, one can draw insights into how systems evolve and behave under varying conditions. For instance, recognizing that certain defects or imperfections can increase the entropy of a crystal can be invaluable in materials science and engineering, while in information theory, gauging the entropy of a message can optimize data storage and transmission protocols. In a broader perspective, the analogous nature of entropy in both fields underscores the universality of certain principles across disparate domains of science. This interconnectedness reveals the beauty of scientific inquiry, where seemingly unrelated concepts converge into universal truths.

Entropy and Philosophy—The Arrow of Time

Exploring the philosophical engagements with entropy touch upon some of the most profound questions about the universe, the nature of time, and the fabric of reality.

The progression from order to disorder, as described by the Second Law of Thermodynamics, offers a potential explanation for the directionality of time. While we perceive time as moving "forward," on the microscopic level, most physical processes are time-reversible. It's the macroscopic increase in entropy that lends a sense of directionality to time, giving rise to what's known as the "arrow of time". Philosophically, this has significant implications. It means our very perception of past, present, and future—concepts so integral to human experience—are bound to this thermodynamic principle. A first question is due: *is "time's arrow" a fundamental aspect of the universe, or is it a mere consequence of increasing entropy?*

In the philosophical realm, however, the concept of entropy prompts several additional questions:

- *Nature of Information: Is information a foundational aspect of reality, akin to space, time, and energy? Or is it merely a derivative concept, emerging from more basic physical laws?*
- *Observer's Role:* The act of gleaning information from a system, and thereby understanding its entropy, necessitates an observer. *Thus, are the observers that define and shape reality?*
- *Physicality of Information:* Given that information and entropy are so closely linked, *does this imply that information has an inherent physicality?* This is a significant question in the philosophy of mind, especially when considering theories that posit consciousness as an information process.

Definitely, entropy, while rooted in physics, has profound philosophical implications. It challenges our perceptions of time, our understanding of causality, and our notions of reality and information. By straddling the realms of physics and philosophy, entropy serves as a poignant reminder of the interconnectedness of the universe and the conceptual frameworks we employ to understand it.

Entropy and Religion—Creation and Decay

Finally, entropy as a scientific concept also provides a unique lens through which we can view various religious and spiritual ideas. Though it is not traditionally a religious concept, its implications can resonate with several religious notions.

Particularly, many religious traditions start with a creation myth where order emerges from chaos. The story of entropy, as a progression from order to disorder, could be seen as mirroring this myth but in reverse. While science describes a universe tending towards disorder, many religions speak of a world that, after its divine creation, becomes tainted by sin, evil, or ignorance, leading to moral and spiritual degradation. For example, in Christianity, the concept of the Fall of Man from the Garden of Eden can be paralleled with a move from order to chaos, or from purity to corruption. Similarly, in Hinduism, the Yugas describe epochs of declining dharma (righteousness), moving from a golden age of order to subsequent ages of increasing moral and societal chaos.

Furthermore, in many religions exists the idea of an "end time" or an eschatological event that signifies a culmination of the world's moral and cosmic decay. This can be seen as a point of maximum entropy, after which a reset or renewal occurs. In Christianity, the idea of the Apocalypse, or in Islam, the Day of Judgement, marks a decisive end to a world that has become morally and spiritually degraded.

Finally, in some Eastern philosophies and spiritual practices, the ultimate goal is the dissolution of the ego, leading to a realization of a more profound, universal truth or unity. This dissolution can be metaphorically related to entropy, where the structured, distinct identity of the ego "degrades" into a boundless, undifferentiated state of consciousness (i.e. a sort of state of maximum entropy). In traditions like Buddhism or certain branches of Hinduism, this might resonate with the concept of attaining Nirvana or Moksha, respectively.

As the concept of entropy could be employed in religious or spiritual discussions, however, it is essential to remember its specific scientific meaning. While entropy offers a compelling metaphor and can generate rich dialogues across various disciplines, its scientific and religious uses are strictly distinct.

The author, here, does not want to oversimplify or misrepresent either domain when drawing parallels. On the contrary, he wants to stimulate thoughts about the interplay of entropy with religious and spiritual concepts underscoring the perennial human effort to find connections between our scientific understanding of the world and our deeper existential and moral inquiries.

References

[1] Starzak, M.E. (2010). Energy and Entropy: Equilibrium to Stationary States. Springer Science & Business Media.

[2] Rajput, R.K. Engineering Thermodynamics, Third Edition. SI Units Version. Engineering Series, Laxmi Publication (P) LTD, Boston, 23, 101-102.

[3] Adamson, A. (2012). A Textbook of Physical Chemistry. Elsevier.

[4] Gray, R.M. (2011). Entropy and Information Theory. Springer Science & Business Media.

[5] Krivovichev, S.V. (2016). Structural complexity and configurational entropy of crystals. Acta Crystallographica Section B: Structural Science, Crystal Engineering and Materials, 72(2), 274-276.

[6] Scott, G.D. & Kilgour, D.M. (1969). The density of random close packing of spheres. Journal of Physics D: Applied Physics, 2(6), 863.

Measuring Entropy

Unlike conserved energies and measurable system variables such as temperature (T) and volume (V), entropy (S) stands apart, bearing unique characteristics.

Though entropy itself isn't directly measurable in the laboratory, the product of temperature and entropy carries energy units. This correlation implies that entropy generated during an irreversible process contributes to the production of energy. Intriguingly, both entropy and its product with temperature correspond to the system's randomness. The relationship between randomness, energy, and entropy unveils one of thermodynamics' captivating facets, and its physical essence can be explored through probabilistic means.

In specific scenarios, such as isothermal and reversible expansions of an ideal gas, the entropy's dependence is solely on the volume change. The equation for this change is represented by:

$$\Delta S = R \ln \frac{V_2}{V_1} \tag{1}$$

Here, the entropy alteration is independent of temperature, relying only on the transformation in the system's volume.

The phenomenon of vaporization presents another intriguing aspect of entropy. As a liquid transitions into a gas, it absorbs heat without a change in the system's temperature. Although energy is introduced into the system, it doesn't manifest as an increase in the molecules' kinetic energy. Rather, the entropy boost during vaporization, expressed as $\Delta S_{vap} = \dfrac{\Delta H_{vap}}{T}$, is linked with the substantial volume alteration in the liquid-to-gas transition. In this instance, the energy infusion serves to elevate the entropy without affecting the system's kinetic energy, illustrating the nuanced relationship between energy, entropy, and the states of matter.

For an isothermal ideal gas, the change in entropy is solely governed by the alteration in the volume of the gas. This change in entropy can be expressed by the equation (1), which is equivalent to the difference:

$$\Delta S = S_2 - S_1 = R \ln V_2 - R \ln V_1 \tag{2}$$

The connection between the entropy of a state and its volume is evident in the direct proportionality to the natural logarithm of the volume, given by:

$$S = R \ln V \tag{3}$$

An intriguing perspective emerges when we consider the entropy at the molecular level. Dividing this entropy by Avogadro's number results in an expression for the entropy per molecule:

$$S' = \frac{S}{N_a} = \frac{R}{N_a} \ln (V) = k \ln (V) \tag{4}$$

where k, having units of $\mathrm{J \cdot molecule^{-1}\ K^{-1}}$, represents Boltzmann's constant.

This molecular-level understanding of entropy maintains the logarithmic dependence on volume. It leads to the fascinating notion that a single molecule can enhance its entropy simply by having the freedom to move within a larger volume. Regardless of its speed, the molecule's precise location becomes increasingly challenging to determine as the volume expands. This concept adds another layer to our understanding of entropy, linking it not just to broad thermodynamic systems but also to the behavior of individual molecules.

Consider a mole of ideal gas expanding isothermally from 1 to 2 liters. This change is characterized by an entropy alteration, represented by:

$$\Delta S = R \ln \frac{2}{1} \tag{5}$$

On a molecular level, the entropy change for one molecule is:

$$\Delta S = k \ln \frac{2}{1} \tag{6}$$

where each molecule initially free to move within 1 liter now has the liberty to move within 2 liters.

Imagine dividing each volume into a set of equal-sized cubes, which can be considered as microvolumes ideally matching the molecular size. A molecule could "visit" any of these cubes with an equal chance. For entropy differences, the ratio of the volumes is what matters. Therefore, as long as the 2-liter volume contains twice the number of cubes as the 1-liter volume, the entropy change can be determined.

These volume cubes define the states of the system. Doubling the volume from 1 to 2 liters translates into doubling the location states.

Interestingly, if we consider two molecules, it might be expected that they would occupy double the number of states. For a 1-liter cube, the expansion to 2 liters yields four distinct states, reflecting different positions of the two particles. For three molecules expanding from 1 to 2 liters, the number of states increases from 1 to $2^3 = 8$. In the case of Avogadro's number of molecules, the states increase in a pattern that aligns with the volume change.

A key principle here is that the difference in entropy corresponds to a ratio of states [1]. Boltzmann formulated an absolute entropy for states as:

$$S = k \ln (f) \tag{7}$$

where $k = 1.38 \cdot 10^{-23}$ J K^{-1} molecule^{-1}.

Though the volume assigned to location states can be arbitrary, definite states can be defined for particular systems. This seminal equation, inscribed on Boltzmann's tombstone in Vienna, signifies a lasting legacy in the field of thermodynamics and statistical mechanics [2].

Boltzmann's interpretation of entropy is founded on the idea that every state within a system is accessible to a molecule and equally probable. Knowing the number of states allows us to determine the probability of finding a molecule in a specific state, which is simply the inverse of the total number of states. For example, in a system with 100 states, the probability of locating a molecule in any particular state is 1/100 or 0.01. By increasing the system's size and states to 200, the probability is halved to 1/200 or 0.005, reflecting the increased randomness and the diminished likelihood of finding the particle in a given state.

In a scenario involving two particles shifting from one to two states, the probability is $1/2^2 = (1/2)^2 = 1/4$, meaning both particles will return to their original state one-fourth of the time.

As the number of particles (N) grows, the probability of returning to the initial state decreases significantly. For instance, with 10 particles expanding from 1 to 2 liters, the return probability is $1/2^{10} = 1/1024 = 0.00098$.

In a system with Avogadro's number of particles, the probability of return becomes so minuscule as to be virtually nonexistent. This extraordinarily low probability underlies the irreversibility of the expansion from 1 to 2 liters; once the gas has expanded, it remains that way.

Such system exhibits what is known as a Poincare cycle [3], where the gas cycles through all possible states before returning to the original state. A system with ten particles and 1024 states will navigate through each of these states before reverting to the initial state. If the average time to transition from one state to the next is known, the Poincare recurrence time simply becomes T × number of states.

Though all states are equally probable in Boltzmann's framework, only one specific state positions all particles on the left, represented as the (L, R) = (N, 0) distribution. Conversely, a balanced 50 – 50 (N/2, N/2) distribution occurs much more frequently, as numerous distinct states contribute to this particular arrangement. This aspect further emphasizes the complexity and intriguing nature of the probability distributions within thermodynamic systems.

Entropy of Mixing and Irreversibility

When two substances such as liquids or gases mix, the process is generally spontaneous and irreversible. The mixed system becomes more disordered or random, resulting in a larger entropy. This process does not naturally reverse, so it is essential to calculate the entropy of mixing in a reversible way.

Consider the irreversible mixing of n_A moles of ideal gas A in a volume V_A and n_B moles of ideal gas B in volume V_B, at a constant temperature and pressure. This mixing can be visualized as joining two cylinders containing these gases. The total volume V_t is then simply:

$$V_t = V_A + V_B \tag{8}$$

The entropy changes when both gases expand during mixing are as follows:

$$\Delta S_A = n_A R \ln\left(\frac{V_t}{V_A}\right) \tag{9}$$

$$\Delta S_B = n_B R \ln\left(\frac{V_t}{V_B}\right) \tag{10}$$

Planck's thought experiment (the so-called Gedanken experiment [4]) allows to calculate the exact entropy change through an ideal reversible process. In this experiment, gases A and B are expanded to V_t in separate cylinders, each capped with special end plates that allow only the other gas to pass through.

The cylinders are then pushed together, forming one larger cylinder with volume V_t, and the gases mix reversibly.

The total entropy for mixing comes from the sum of several reversible steps:

I.   A and B are expanded to volume V_t in separate cylinders.
II. The reversible mixing step involves no change in internal energy (temperature constant) and no work ($P_{ext} = 0$), so that $\Delta E = 0 = q + w = q = q_{rev}$.

Given these conditions, the total entropy of mixing in the Gedanken experiment is:

$$\Delta S = n_A R \ln\left(\frac{V_t}{V_A}\right) + n_B R \ln\left(\frac{V_t}{V_B}\right) + 0 \tag{11}$$

We can further express the entropy in terms of mole fractions. Given the total moles $n_t = n_A + n_B$, we have the following ratios:

$$\frac{V_t}{V_A} = \frac{n_t}{n_A} \tag{12}$$

$$\frac{V_t}{V_B} = \frac{n_t}{n_B} \tag{13}$$

Substituting these mole fractions, the total entropy of mixing becomes:

$$\Delta S_{mix} = -R\left[n_A \ln(X_A) + n_b \ln(X_B)\right] \tag{14}$$

Expressing Equation (14) in terms of the total moles, it is possible to find the entropy per total moles:

$$\overline{\Delta S_{mix}} = -R\left[X_A \ln(X_A) + X_b \ln(X_B)\right] \tag{15}$$

Equation (15) can be generalized for a mixture of N species with mole fractions X_i:

$$\overline{\Delta S_{mix}} = -R\sum_{i=1}^{N} X_i \ln(X_i) \tag{16}$$

Definitely, Equation (16) captures the entropy changes associated with mixing in a reversible context, offering a robust mathematical description of this naturally irreversible process.

Molar Entropy to Single Molecule Entropy

The molar entropy of mixing, given by Equation (16), can be converted to an entropy of mixing for a single molecule by dividing both sides by Avogadro's number (N_A):

$$\Delta S = -k\sum_{i=1}^{N} X_i \ln(X_i) \tag{17}$$

Where $k = \dfrac{R}{N_A}$ is the Boltzmann constant.

Now, the following question may arise: "How can there be an entropy of mixing for one single molecule?"

The answer to such a question lies in interpreting the mole fraction as a probability. In fact, for a large number of particles, the mole fraction for A is equivalent to the probability that a molecule A is selected if one particle is randomly pulled from the mixture.

For one particle that is either A or B, $X_A = p_A$ and $X_B = p_B$. Considering the case when both A and B are equally probable, $p_A = p_B = 0.5$; then the entropy of mixing in such system is:

$$\Delta S_{mix} = -k\left[0.5\ln(0.5) + 0.5\ln(0.5)\right] = k\ln 2 \approx k \cdot 0.69 \tag{18}$$

This is the largest possible entropy for a two-species mixture. Since both options are equally probable, the choice is most random.

Conversely, considering the case when the probabilities are different and, for example: $p_A = 0.8$ and $p_B = 0.2$; then the entropy of mixing in such system is:

$$\Delta S_{mix} = -k\left[0.8\ln(0.8) + 0.2\ln(0.2)\right] \approx 0.5k \tag{19}$$

Here, the order is reflected in lower entropy, consistent with the fact that the system is less random.

The famous Boltzmann equation [2] is then derived from the entropy of mixing when W identical states are mixed, i.e. the probability of each state is $1/W$ and the entropy of mixing is:

$$\Delta S_{mix} = -k \sum_{i=1}^{W} \frac{1}{W} \ln\left(\frac{1}{W}\right) = k \ln W \tag{20}$$

The connection between mole fractions and probabilities enables the calculation of the entropy of mixing for single molecules. The more random the system (i.e. the closer the mole fractions are to equal), the higher the entropy. This approach ties in with the foundational principles of statistical mechanics and provides insights into the microscopic behavior of systems.

Entropy of Mixing and Reference State

The entropy of mixing is relative to a starting entropy of zero when A and B are located exactly in their respective containers. Since $\ln(1) = 0$, when assigning a probability of 1 for the unmixed state, the entropy difference stems from the change to the mixed state.

From this point of view, Boltzmann's equation, i.e. Equation (20), defines the absolute entropy of a system by referring to a specific reference state where the system is in one unique state, and thus has an entropy of 0. From this concept, we can derive the third law of thermodynamics, in terms of stating that the entropy of a perfect crystalline substance at the absolute zero temperature (0 K) is zero. In other words, at 0 K:

I.   Kinetic energy of each classical atom is 0.
II. Atoms are motionless within their cavities.
III. The entire crystal is in one single state.

Thus, the entropy for a perfect crystal at 0 K is:

$$S = k \ln 1 = 0 \tag{21}$$

As the temperature or volume of a crystal increases, the corresponding entropy change becomes an absolute entropy since its reference entropy is 0 at $T = 0$ K:

$$\Delta S = S_f - S_i(T = 0K) = S(T,V) - 0 \tag{22}$$

In other words, the calculation of the absolute entropy at various temperatures involves summing all entropy changes starting from the absolute zero ($T = 0K$).

However, some systems, like certain crystals and molecules, may exhibit a residual entropy at 0 K. This residual entropy arises from the distinguishable microstates that can exist at absolute zero.

For example, carbon monoxide (CO) molecule demonstrates residual entropy, as C and O atoms in CO are roughly the same size, and a CO molecule can fit into its lattice site as either CO or OC. The two orientations result in two equally probable states at 0 K. For N such sites (or molecules), there are 2^N distinct perfect crystals for CO, leading to the following absolute entropy at 0 K for CO:

$$S = k \ln(2^N) = R \ln 2 \tag{23}$$

Conversely, symmetrical linear molecules like O_2 and CO_2 do not have residual entropy because quantum mechanics treats two like atoms as indistinguishable. Without labels, two orientations become indistinguishable, thus eliminating residual entropy.

The example of CO (as well as many other molecules could do, as crystalline ice too) underscores the complex and nuanced nature of statistical thermodynamics and, in particular, of entropy, reflecting how detailed knowledge of a system's microstates can provide a deeper understanding of its macroscopic properties.

The absolute entropy of the elements in their most stable states at 25°C and 1 atm is determined through these calculations. Tables of absolute entropies are often listed at this standard temperature and pressure to aid in thermodynamic calculations, such as Gibbs free energy, enthalpy, and internal energy changes [5, 6].

Hydrophobia

As a final remark of this chapter, the so-called hydrophobic effect, an intricate and still-incomplete understood entropic force [7], will be discussed to further highlight the complexity of entropy and its correlation with the microscopic world.

Apolar substances like oils, organic solvents such as alkanes, and benzene are known not to mingle with water. For instance, oil and water, when shaken, quickly segregate into two layers. This phenomenon, where apolar solutes dissolve poorly in water or even aggregate and fall out of the solution if they surpass their solubility threshold, is called the hydrophobic effect.

The counterintuitive hydrophobic effect mainly arises from its dependency on (liquid) water's properties, a rather unique solvent. Even though the energy levels involved in the hydrophobic effect are fairly understood [8, 9], many aspects of it are still unclear.

Form this chapter's point of view, a well-mixed solution of oil and water should theoretically reach a state of maximum entropy, making solvent and solute indistinguishable, but instead, demixing occurs, apparently reducing entropy. This reduction, however, is only superficial. Thermodynamic studies [10, 11] have revealed that the water surrounding apolar solute possesses lower entropy compared to the rest, and demixing actually leads to an increase in overall entropy.

The prevailing literature theory is that the water surrounding an apolar solute is compelled to create clathrate-like formations. Characterized by pentagonal rings of hydrogen-bonded water, clathrates are more structured than bulk water, with their positions and orientations more confined, thus diminishing their spatial entropy.

According to this hypothesis, bringing two separated apolar solute molecules close enough to remove water from their interface should enhance the overall entropy since the quantity of water in contact with the apolar surface decreases. At some point in this transition, a positive gradient in entropy appears, creating an attractive hydrophobic force.

This significant hydrophobic effect holds major implications in biology, influencing the structure and stability of proteins, and encouraging the self-assembly of biological membranes.

A similar (apparently) counterintuitive effect of reducing entropy could be attributed to the formation of the high-entropy oxides, in which a simple and well-ordered crystalline system is favored over multiple crystalline systems under certain conditions (thoroughly discussed in the next chapters). Anyway, as the hydrophobic effect, the high-entropy materials' stabilization too is always justified by an increase of the overall entropy of a given system, in accordance to the *ever-right* second law of thermodynamics.

References

[1]   Thess, A. The Entropy Principle: Thermodynamics for the Unsatisfied. Springer.
[2]   Johnson, E. (2018). Anxiety and the Equation: Understanding Boltzmann's Entropy. MIT Press.
[3]   Tipler, F.J. (1979). General relativity, thermodynamics, and the Poincaré cycle. Nature, 280(5719), 203-205.
[4]   Norwich, K.H. (2016). Boltzmann–Shannon entropy and the double-slit experiment. Physica A: Statistical Mechanics and its Applications, 462, 141-149.
[5]   Jenkins, H.D.B. & Glasser, L. (2003). Standard absolute entropy, values from volume or density. 1. Inorganic Materials. Inorganic Chemistry, 42(26), 8702-8708.
[6]   Glasser, L. & Jenkins, H.D.B. (2004). Standard absolute entropies, S 298, from volume or density: Part II. Organic Liquids and Solids. Thermochimica Acta, 414(2), 125-130.
[7]   Sharp, K. (2019). Entropy and the Tao of Counting: A Brief Introduction to Statistical Mechanics and the Second Law of Thermodynamics. Springer Nature.
[8]   Tanford, C. (1980). The Hydrophobic Effect: Formation of Micelles and Biological Membranes, 2d ed. J. Wiley.
[9]   Ben-Naim, A.Y. (2013). Solvation Thermodynamics. Springer Science & Business Media.
[10] Kronberg, B., Costas, M. & Silveston, R. (1995). Thermodynamics of the hydrophobic effect in surfactant solutions: Micellization and adsorption. Pure and Applied Chemistry, 67(6), 897-902.
[11] Del Rio, J.M. & Jones, M.N. (2001). Thermodynamics of the hydrophobic effect. The Journal of Physical Chemistry B, 105(6), 1200-1211.

Information Theory (Shannon Entropy) and Materials Science

In modern times, the most prevalent method to gauge the quantity of data in a symbolic message is by computing the 'information entropy', a concept put forth by Claude Shannon in 1948 [1]. Yet, we're not just limited to textual segments when referring to messages; other entities can be deemed messages too. For determining the information entropy of a symbol in any message, the only prerequisite is being aware of the likelihood of these symbols occurring within the message.

Particularly, Shannon devised a method to calculate the quantity of information, denoted as H, within a message composed of P symbols distributed across N equivalence categories. This is articulated as:

$$H_n(P) = H_n(p_1, p_2, \ldots, p_N) = \sum_{i=1}^{N} L(p_i) \tag{1}$$

Where, p_i denotes the occurrence probability of the ith symbol, calculated as the ratio $p_i = P_i/P$.

These N individual probabilities p_i constitute a discrete probability distribution, where $N = |P|$.

It's worth noting that these probabilities can represent a multiset; allowing for repeated elements. Rearranging them doesn't affect Shannon entropy (due to its symmetry). Including or excluding probabilities where $p_i = 0$ also doesn't impact it (owing to its expansibility). Furthermore, two conditions persist:

   i. $0 \leq p_i \leq 1$;

   ii. $\sum_{i=1}^{N} p_i = 1$.

Condition (ii) ensures a complete probability distribution. Nevertheless, variations of Shannon entropy do exist that cater to incomplete probability distributions.

From a foundational perspective in information theory, Shannon entropy emerges as the quintessential metric of information. This is due to its alignment with numerous intuitive mathematical properties anticipated from such a measure [2].

Recently, Krivovichev [3] adapted Shannon's theory to evaluate crystal structures, interpreting the sequence of orbits (or systems of positions that are symmetrically equivalent) filled by atoms as the message, with the atoms symbolizing the symbols. Therefore, the structural data an individual atom of a crystal structure holds can be represented as follows:

$$I_G^{str} = -\sum_{i=1}^{k} p_i \log_2 p_i \ (\text{bits/atom}) \tag{2}$$

$$p_i = m_i \big/ v \tag{3}$$

Here, m_i denotes the multiplicity of the ith crystallographic orbit, v denotes the atom count in the simplified unit cell, and k stands for the number of crystallographic orbits in the crystal's structure. It's vital to ensure the cell is meticulously minimized since merely replicating the structural design through translations doesn't genuinely introduce new data. This structural intricacy measure also defines the quotient graph's complexity of the structure. In essence, this is the finite graph comprising all v vertices of the pared-down unit cell of the crystal structure, in addition to the edges linking the respective vertices in the crystal structure. Edges that connect translationally identical vertices are transformed into loops [4].

For example, when considering the quotient graphs of the crystal structures of α- and β-quartz, they are indistinguishable. In both configurations, both the Si and O atoms occupy a single orbit. Specifically, their multiplicities are 3 for Si and 6 for O. As a result, the calculated structural I_G^{str} is given by:

$$I_G^{str} = -\left[\frac{3}{9}\cdot\log 2\left(\frac{3}{9}\right) + \frac{6}{9}\cdot\log 2\left(\frac{6}{9}\right)\right] \approx 0.918\,\text{bits/atom} \tag{4}$$

Several additional complexity metrics can be derived based on the foundational principles of information content in crystal structures:

- **Maximal Information Content**, i.e. the highest information content of a crystal structure, when expressed in bits per atom, defined as:

$$I_{G,\,max}^{str} = \log_2 M \tag{5}$$

where M is the total number of atoms of a given structure. This condition represents the case of evenly distributed probabilities (i.e. $p_i = 1/M$). Within the realm of information theory, this peak Shannon entropy is occasionally referred to as Hartley entropy.

- **Normal Information Content**, representing the relative measure of information content, derived as:

$$I^{str}_{G,norm} = I^{str}_G \Big/ I^{str}_{G,max} \tag{6}$$

The normal information content is a dimensionless value that spans between 0 (when all atoms are symmetrically identical) and 1 (where no atoms are symmetrically identical).

As an example, by applying Equation (2) and Equation (6) to fluorite and bixbyite crystal structures, being very similar but owning different symmetries, the following values are obtained: $I^{fluorite}_{G,norm} = 0.256$ and $I^{fluorite}_{G,norm} = 0.205$. A heightened normal information content in the fluorite structure can be attributed to its elevated symmetry [5]. This can be likened to an unbiased dice roll where the likelihood of a particular cation occupying a cell spot is evenly distributed. On the contrary, the bixbyite structure, which has less symmetry compared to the fluorite structure, can be equated to a biased dice roll. In this case, specific sites within the structure have a higher probability of being filled by certain cations over others. Essentially, while the fluorite structure provides an equal probability for cation placements, the bixbyite structure shows a preferential positioning for some cations.

- **Total Information Content**, articulated as:

$$I^{str}_{G,tot} = -v \sum_{i=1}^{k} p_i \log_2 p_i \text{ (bits/unit cell)} \tag{7}$$

Taking again as an example both α- and β-quartz crystal structures, they exhibit an $I^{str}_{G,tot}$ value of approximately 8.265 bits/unit cell.

Extending the $I^{str}_{G,tot}$ metric, all the inorganic substance crystal structures can be categorized into five classes [6]:

i.   very simple (< 20 bits/unit cell);
ii. simple ($20 - 100$ bits/unit cell);
iii. intermediate ($100 - 500$ bits/unit cell);
iv. complex ($500 - 1000$ bits/unit cell);
v.   highly complex (> 1000 bits/unit cell).

Notably, the crystal structures of materials like zeolites and other microporous frameworks frequently fall into the "highly complex" category. Ewingite's framework structure, with the chemical formula $Mg_8Ca_8[(UO_2)_{24}(CO_3)_{30}O_4(OH)_{12}(H_2O)_8](H_2O)_{130}$, stands out as the most intricate mineral structure ever discovered, boasting an $I^{str}_{G,tot}$ value of a staggering 23,477.507 bits/unit cell [7].

A noteworthy aspect is that all the previously presented complexity measures are agnostic to the specific metrics of a given crystal structure. In fact, their foundation is rooted in combinatorial principles. That said, it's feasible to define an 'information density' for a given crystal structure, denoted as:

$$\rho^{str}_{inf} = I^{str}_{G,total} \Big/ V^{str}_{red} \text{ (bits/Å}^3) \tag{8}$$

where V_{red}^{str} represents the volume of the reduced unit cell of a given structure.

Delving more into the thermodynamic entropy of a crystal, it encompasses components such as configurational, vibrational, and other minor types of entropy. Configurational entropy (S_{conf}) is related to the spatial arrangement of atoms which is temperature independent, whilst vibrational entropy (S_{vib}) is related to a temperature-dependent contribution describing the movement of atoms which is itself tied to the atomic interactions [8].

The vibrational component has attracted considerable attention, leading to the development of numerous methodologies for its precise computation [9, 10]. In essence, to accurately measure the vibrational entropy of a certain system, meticulous multi-temperature, high-resolution single-crystal X-ray diffraction analyses are imperative. These are essential for refining Anisotropic Displacement Parameters (ADPs) of all atoms, notably hydrogen atoms, which are notoriously tricky to hone from X-ray diffraction data. Additionally, periodic DFT calculations serve as invaluable tools in this context [11]. Nonetheless, it's essential to bear in mind the myriad complexities surrounding ADPs. They often serve as catch-alls for all forms of experimental inconsistencies. Factors such as inadequate absorption correction, partial data, minor disorder presence, different scattering types, twinning, challenges in thermal motion and charge density deconvolution, can markedly influence the ADPs. Yet, even considering perfect ADPs, it is not a given that a compound with more expansive ADPs would inherently possess heightened entropy. This nuance becomes even more pronounced when observing conformational polymorphs [11].

The study of configurational entropy has largely remained in the shadows, with the notable exceptions being investigations into mixing entropy and order–disorder phase transitions [12-14]. Broadly, the configurational component of entropy refers to the segment of total entropy exclusively derived from the spatial arrangement of its constituents, devoid of influences from their dynamical behaviors. Recent advancements have enabled *in situ* direct measurements of the vibrational spectra of both strong and fragile metallic phases across various states, such as glasses, liquids, and crystals [15]. Interestingly, for both strong and fragile variants, the vibrational entropies of the liquid and glass states displayed only a minimal increase over their crystalline counterparts, making up less than 5% of the entire surplus entropy as derived from step calorimetry. This finding underscores that the surplus entropy of metals is predominantly configurational in nature [16].

A recent research has unveiled that a crystal structure's reduction in configurational entropy relative to its maximum attainable value corresponds linearly to I_G^{str} [8]. More in detail, the information content as provided by crystallographic data contributes negatively to the configurational entropy of a given structure:

$$S_{conf} = S_{conf\_max} - I_G^{str} k_B N \ln(2) \tag{9}$$

where S_{conf} is the configurational entropy, S_{conf_max} is the maximum configurational entropy obtained when all atoms (positions) are symmetrically equivalent, I_G^{str} refers to the one obtainable in Equation (2), N the number of atoms in the crystal, k_B is the Stephan-Boltzmann constant and ln (2) is a conversion factor between binary and natural logarithms.

The concept of Equation (9) can be perceived as a quantitative affirmation of the intuitive notion that adding complexity to a crystal structure inversely affects its configurational entropy.

To broaden the entropy-related complexity measures introduced by Krivovichev works [3, 6, 16], Hornfeck [17] pinpointed some additional Quantitative Crystal Structure Descriptors (QCSDs) [18, 19]. Specifically, QCSDs are integer in nature, and their ratios with their total values in the denominator are conducive to defining a probability distribution and the accompanying entropy measures.

Generally speaking, a crystal structure's comprehensive geometric characterization includes three facets: (i) its symmetry, as designated by the space-group type; (ii) its metrics, as outlined by the lattice parameters; and (iii) its atomic coordinates, as discerned for an asymmetrical unit, specified through the occupied Wyckoff positions.

Within a particular space-group type and setting, a specific Wyckoff letter (ranging from a to z, including α) can identify each Wyckoff position. This set of letters constitutes a complete alphabet, though it is only fully present for the space-group type *Pmmm*, No. 47. This means a crystal structure's geometry can be succinctly represented through linear encoding, i.e. a crystal structure Wyckoff sequence. This sequence comprises the space-group type number followed by a collection of Wyckoff letters. These letters are arranged in reverse alphabetical order, with each letter's frequency of occurrence denoted as a superscript.

Thus, relying on standardized crystal structure data, a Wyckoff sequence serves as a distinctive encoding for an abstract crystal structure. It is worth noting that the number of unique Wyckoff sequences is relatively small, with just 15,503 instances found in the Pearson's Crystal Data Crystal Structure Database for Inorganic Compounds [20].

Most parameters utilized in the depiction of crystal structures are inherently numerical.

First of all, a straightforward solution to designating atom types is to employ the atomic number Z_i of the chemical element in the structure, with i representing the corresponding element in the probability distribution Z.

Furthermore, the 'site arity' is employed as descriptor for the degrees of freedom associated with a given Wyckoff position concerning its general atomic coordinates, as recorded in the International Tables for Crystallography, Vol. A [21].

The crystallographic scenarios are also captured by descriptors such as invariant, univariant, bivariant, and trivariant, which correspond to a fixed

Wyckoff position with zero, one, two, or three degrees of freedom, respectively. In mathematical terms, these descriptors can be referred to as constant, univariate, bivariate, and trivariate. The overarching concept of symmetry of a generic crystal structure is then captured by the division pattern of the multiplicities of the crystallographic orbits.

The choice to represent the chemical degrees of freedom in a crystal structure using the atomic number of its constituent elements is essential to translate the qualitative information about the atom type into a quantitative parameter. Even though Z_i is not the only possible choice to do that, it is the most meaningful to distinguish different systems owning the same crystal structure. As an example to that, Hornfeck [17] consider the following three compounds: NaCl, KCl, and RbCl, all having the Wyckoff sequence 225, *ba*. Looking at their chemical complexity, from the point of view of their valence electrons (determining their chemistry) they do not differ. However, considering their total electron distributions, they do. The total electrons for the ion pairs (M^+, X^-) are (10, 18) for NaCl, (18, 18) for KCl, and (36, 18) for RbCl. This leads to corresponding entropies calculated through Equation (1) of:

- H(10, 18) = 0.940 (NaCl);
- H(18, 18) = 1.000 (KCl);
- H(36,18) = 0.918 (RbCl);

with the related Shannon entropy being at its peak for the equidistributed case (KCl).

To summarize the above descriptors to numerically depict the various existing crystal structures, there are three integer crystal structure descriptors that can be used to define three corresponding sets of information measures. These descriptors are:

   i. the atomic numbers of the constituting atoms (Z_i);
  ii. the site multiplicities (M_i);
 iii. the site arities (A_i).

Such descriptors align with Krivovichev's approach [refs] and can be formulated into three tetrads, as shown in the following equation:

$$\mathfrak{I}_X = \{I_X, I_{X,max}, I_{X,norm}, I_{X,tot}\}$$

where the subscript X denotes the type of attribute, with X belonging to the set $\{Z, M, A\}$.

Additionally, Hornfeck [17] introduced a unique notation for I_X, denoting the chemical (I_Z), combinatorial (I_M), and coordinational (I_A) complexity of a given crystal structure.

Figure 1 [17] provides a visual illustration of these three conceptually distinct contributions to a crystal structure's entropy/complexity. In particular, Figure 1 reveals how the complexity emerges from subdividing a crystal structure

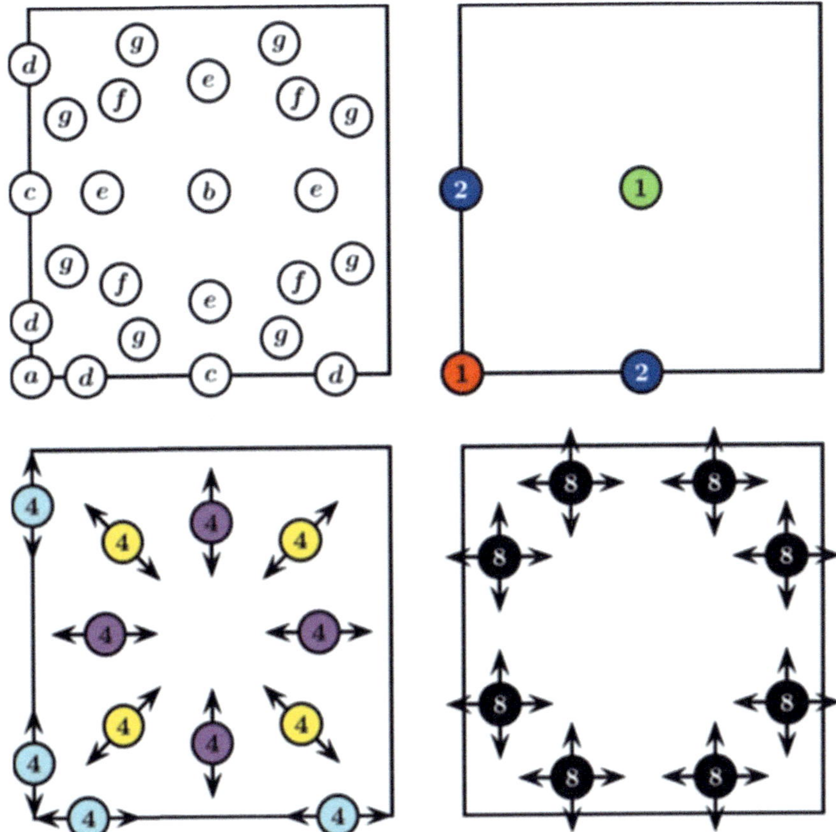

Fig. 1: The Wyckoff positions in plane group *p4mm* (No. 11), (top left) designated by their Wyckoff letters, and shown separately for positions with (top right) zero, (bottom left) one and (bottom right) two degrees of freedom (arities) A_i, as indicated by the presence of zero, one and two (orthogonal) pairs of arrows, respectively [17].

composed of atoms, each with Z_i degrees of freedom, into Wyckoff positions, each contributing the pair (M_i, A_i) degrees of freedom.

Very recently [8], an open source Python-based program called *crystIT* has been developed for calculating the information content of crystal structures based on the theoretical concepts and equations presented in this chapter. The source code of *crystIT* is provided as a ready-to-use Python file, freely available at https://github.com/GKieslich/crystIT.

Definitely, the application of the information theory concepts, derived from the Shannon Entropy concept, made by Krivovichev [3, 6, 16] and refined by Hornfeck [17] offers a systematic way to describe the complexity of a crystal structure from different perspectives. However, further progress is still needed in this entropy-related research field, particularly to bring the concept of Shannon entropy closer to applied materials science.

Possible future research directions in this field could include quantitative analyses of calorimetric data to better understand phase-transition thermodynamics in inorganic materials and formulating correlations and/or predictors for stable high-entropy materials.

References

[1] Shannon, C.E. (1948). A mathematical theory of communication. The Bell System Technical Journal, 27(3), 379-423.

[2] Aczél, J. & Daróczy, Z. (1975). On measures of information and their characterizations. Mathematics in Science and Engineering, 115, 234. ISBN: 0120437600. Academic Press.

[3] Krivovichev, S. (2012). Topological complexity of crystal structures: Quantitative approach. Acta Crystallographica Section A: Foundations of Crystallography, 68(3), 393-398.

[4] Klee, W.E. (2004). Crystallographic nets and their quotient graphs. Crystal Research and Technology. Journal of Experimental and Industrial Crystallography, 39(11), 959-968.

[5] Spiridigliozzi, L., Ferone, C., Cioffi, R. & Dell'Agli, G. (2021). A simple and effective predictor to design novel fluorite-structured High Entropy Oxides (HEOs). Acta Materialia, 202, 181-189.

[6] Krivovichev, S.V. (2014). Which inorganic structures are the most complex? Angewandte Chemie International Edition, 53(3), 654-661.

[7] Krivovichev, S.V. (2020). Polyoxometalate clusters in minerals: Review and complexity analysis. Acta Crystallographica Section B: Structural Science, Crystal Engineering and Materials, 76(4), 618-629.

[8] Kaußler, C. & Kieslich, G. (2021). crystIT: Complexity and configurational entropy of crystal structures via information theory. Journal of Applied Crystallography, 54(1), 306-316.

[9] Fultz, B. (2010). Vibrational thermodynamics of materials. Progress in Materials Science, 55(4), 247-352.

[10] Benisek, A., Dachs, E., Salihović, M., Paunovic, A. & Maier, M.E. (2014). The vibrational and configurational entropy of α-brass. The Journal of Chemical Thermodynamics, 71, 126-132.

[11] Kofoed, P.M., Hoser, A.A., Diness, F., Capelli, S.C. & Madsen, A.Ø. (2019). X-ray diffraction data as a source of the vibrational free-energy contribution in polymorphic systems. IUCrJ, 6(4), 558-571.

[12] Saito, K. & Yamamura, Y. (2005). Configurational entropy and possible plateau smaller than R ln 2 in complex crystals. Thermochimica Acta, 431(1-2), 21-23.

[13] Benisek, A. & Dachs, E. (2012). A relationship to estimate the excess entropy of mixing: Application in silicate solid solutions and binary alloys. Journal of Alloys and Compounds, 527, 127-131.

[14] Benisek, A. & Dachs, E. (2015). The vibrational and configurational entropy of disordering in Cu3Au. Journal of Alloys and Compounds, 632, 585-590.

[15] Smith, H.L., Li, C.W., Hoff, A., Garrett, G.R., Kim, D.S., Yang, F.C. & Fultz, B. (2017). Separating the configurational and vibrational entropy contributions in metallic glasses. Nature Physics, 13(9), 900-905.

[16] Banaru, A.M., Aksenov, S.M. & Krivovichev, S.V. (2021). Complexity parameters for molecular solids. Symmetry, 13(8), 1399.

[17] Hornfeck, W. (2020). On an extension of Krivovichev's complexity measures. Acta Crystallographica Section A: Foundations and Advances, 76(4), 534-548.

[18] Mackay, A.L. (1984). Descriptors for complex inorganic structures. Croatica Chemica Acta, 57(4), 725-736.

[19] Hornfeck, W. (2012). Quantitative crystal structure descriptors from multiplicative congruential generators. Acta Crystallographica Section A: Foundations of Crystallography, 68(2), 167-180.

[20] Villars, P. & Cenzual, K. (2007). Pearson's Crystal Data: Crystal Structure Database for Inorganic Compounds. ASM International, Materials Park, 2010.

[21] Aroyo, M.I. (2013). International Tables for Crystallography. John Wiley and Sons Limited.

High-Entropy Materials—Definition and Thermodynamics

The study of metal oxides' functional attributes spans a broad spectrum of essential and practical scientific domains, including materials science, chemistry, physics, and engineering. To distill this complexity, there are generally two primary methods to customize these functional characteristics. One approach involves chemically altering the oxide through the substitution of its native cations with other selectively chosen ones, based on criteria such as electronic structure, ionic size, and additional fundamental characteristics. The other method focuses on selecting the most suitable structure, morphology and/or micro/nanostructure, achievable through different fabrication processes. Oversimplifying, material control hinges on the constituent elements and how their immediate atomic surroundings affect electron interactions.

However, in practical terms, creating a material with specific functional features is a complex endeavor, as several factors like chemical composition, form, microstructure, stability, functional attributes, and synthesis methodology are interconnected. Additionally, thermodynamic constraints upon synthesis and/or fabrication process may restrict the realization of the desired features. Leveraging configurational entropy as a thermodynamic incentive for stable solutions unveils an expansive, yet largely unexplored, realm of materials that cannot be accessed through methods stabilized by enthalpy alone.

As seen in previous chapters, the concept of ideal entropy is viewed as the entropy value when a system reaches maximum disorder; as order increases, the entropy diminishes from this maximum value. From a conceptual standpoint, a system's entropy serves as a *reservoir* where entropy can either be added to randomize structure or removed to stabilize materials featuring chemical, structural, or electrical ordering within a high-entropy matrix. In other words, the strategic use of entropy in designing novel materials for specific applications aims to create unique crystalline structures, oxidation states, coordination numbers, and lattice distortions.

Generally speaking, high-entropy materials are unique in displaying long-term structural crystalline order coupled with short-term compositional randomness.

This short-term disorder, stemming from the compositional complexities, has an impact on various material properties like thermal, thermoelectric, dielectric, and magnetic ones. These are contingent on several elemental features like valence states, electronegativities, preferred coordination numbers, ionic sizes, and atomic masses, and can be manipulated only to a limited extent. Furthermore, this short-term disorder coexists with short-term order, as seen in both high-entropy alloys featuring metal–metal bonds [1] and oxides with ionic or covalent bonds [2]. This raises questions about how to accurately define the "level of entropy" in a given material composition, from both theoretical and experimental perspectives.

Determining the appropriate criteria for classifying a material as a "high entropy" oxide is a complex task. Existing literature often overlooks the calculation of configurational entropy when identifying a substance as a high entropy oxide. Frequently, the term is loosely applied to any material that contains five or more equimolar components. Moreover, there is also ambiguity in distinguishing between materials based on their different levels of configurational entropy. Terminology is not consistently applied, leading to confusion among commonly used phrases like "compositionally complex", "high entropy", and "entropy stabilized".

To address these issues, Brahlek et al. [3] proposed a hierarchical classification scheme that is both exhaustive and easy to understand (see Fig. 1). Although the focus of such classification is on oxides, that framework is applicable to all materials that possess both cationic and anionic sublattices, including other types of high entropy ceramics such as carbides, nitrides, borides and silicides.

In this scheme, materials containing multiple elements across one or more sublattices are termed "Compositionally Complex Materials" (CCM). "High Entropy Oxides" (HEO) are defined as materials in which configurational entropy contributes to formation and stability but is not necessarily the dominating factor

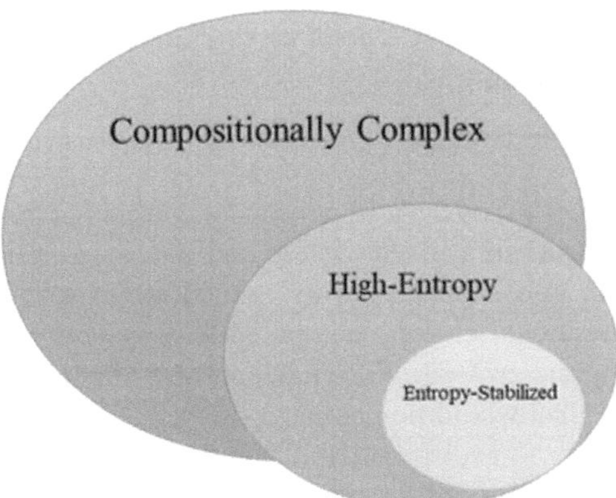

Fig. 1: Schematics of the terminology proposed by Brahlek et al. [3]. Note that all Entropy-Stabilized materials are High-Entropy materials, but not all High-Entropy materials are Compositionally Complex.

(i.e. native defects may also play a role in stabilization). Finally, the term "Entropy Stabilized Oxides" (ESO) represents a more narrowly defined subgroup of high entropy materials, where entropy is the overriding term affecting the formation of a specific crystalline phase that some parent materials may not exhibit.

From a thermodynamic point of view, an entropy-based classification for highly-disordered materials, based on configurational entropy (ΔS_{conf}) calculated using Boltzmann's equation (Chapter 1, Equation (1)) have been proposed:

- High-Entropy Materials (HEMs) having $\Delta S_{conf} \geq 1.5R$
- Medium-Entropy Materials (MEMs) having $R \leq \Delta S_{conf} \leq 1.5R$
- Low-Entropy Materials (LEMs) having $\Delta S_{conf} < R$

When taking temperature and pressure as the system's independent variables, the change in configurational entropy (ΔS_{conf}) influences the entropic component of the Gibbs-Helmholtz equation:

$$\Delta G_{mix} = \Delta H_{mix} - T\Delta S_{mix} = \sum_{i \neq j=1} \beta_{i,j} x_i x_j - T\Delta S_{mix} \tag{1}$$

Here, $b_{i,j}$ denotes the molar constant specific to the involved species, and x_i and x_j denote their mole fractions. ΔG_{mix} must always be negative to let phase formation, while the term $|T\Delta S_{mix}|$ is inherently positive. This allows for the identification of three scenarios concerning the enthalpic term:

- Regular: $\Delta S_{mix} \neq 0$ and $\Delta G_{mix} = -\Delta H_{mix} - T\Delta S_{mix}$;
- Quasi-Chemical: $\Delta S_{mix} \neq 0$ and $\Delta G_{mix} = \Delta H_{mix} - T\Delta S_{mix}$;
- Ideal: $\Delta S_{mix} = 0$ and $\Delta G_{mix} = -T\Delta S_{mix}$.

As temperature increases, the system's reliance on the entropic term for Gibbs free energy also heightens. This creates optimal conditions for forming a stable, single-phase solid solution. The enthalpic term's magnitude is crucial for phase stability; if it closely matches the entropic term, the system will not be stable as a high-entropy material (HEM). Therefore, the thermodynamic condition for a generic HEM phase stability is:

$$|T\Delta S_{mix}| > |\Delta H_{mix}| \tag{2}$$

Recent researches [4, 5] have established a correlation between the number of constituent elements in a given system and its configurational entropy. For instance, a system constituted by only two different elements has a lower maximum configurational entropy compared to a system constituted by three different elements, and so forth. For High-Entropy Materials, fewer crystalline phases are formed than would be predicted by the 'Gibbs phase rule', and they often exhibit uniform cation distribution within the crystal lattice, thus eliminating local clustering of elements and resulting in high configurational entropy values.

Therefore, the complexity in a material composition contribute to an elevated configurational entropy. At high temperatures, this can facilitate the minimization of Gibbs free energy. In the case of entropy-stabilized oxides (ESOs), it can be a primary factor in forming a stable and homogenous phase. However, high configurational entropy is not the sole determinant of a material's functional properties. In fact, it is more accurate to say that the combination of compositional complexity and localized interactions, arising from short-range disorder, defines these properties. High configurational entropy serves merely as an indicator of the extent of this disorder.

In materials like oxides, where the anion sublattice is chemically consistent, local disorder tends to be lower than in alloys. Assuming a generic MO oxide has its M site filled with multiple types of cations, the disorder manifests in variations in bond lengths, angles, and energies within the O–M–O–M framework. These variations significantly influence the material's functional attributes, whether they relate to nearest-neighbor interactions, such as magnetic properties, or to charge and energy transport mechanisms like conductivity and catalytic activity. Configurational entropy also affects high-temperature stability and other temperature-sensitive properties in ESOs. Material functionality is largely determined by the local composition of microstates, which correlates with configurational entropy.

Two key variables influence the behavior of compositionally complex materials (CCMs): the types of elements present and their mixing ratios.

In ionic conduction, for instance, thermally-activated ion movement is sensitive to compositional complexity. The presence of specific elements, like acceptor-type constituents in a perovskite, influences the number of oxygen vacancies, affecting ion mobility. Other factors, such as lattice symmetry and unit cell dimensions, also play a role. Generally, shorter hopping distances and higher lattice symmetries are beneficial as they lower the required activation energy for ion movement. The simplest example in conventional materials is doped ceria ($Ce_{1-x}M_xO_{2-\delta}$), where the degree of local disorder can affect both the pre-exponential term and the activation energy in ionic conduction. Yashima and Takizawa [6] found that doped ceria exhibits higher static and dynamic positional disorder compared to pure ceria, leading to increased ionic conductivity but also to a rise in activation energy due to more diverse energy barriers.

Anyway, different materials respond differently to compositional complexity and short-range disorder, which may lead to unpredicted improvements in functionality. For example, lithium-conducting oxides with complex compositions exhibited radically different behaviors [7-9]. A recent study by Wang et al. [10] partially shed light on the synergistic interaction among multiple cations in lithium-conducting high-entropy oxides (HEOs) at both atomic and nanoscopic levels during electrochemical processes. Specifically, elements with higher electronegativity construct a three-dimensional metallic nano-grid that is electrochemically inert, thereby facilitating electron movement. Meanwhile, the electrochemically inactive cations help stabilize an oxide nanophase. This nanophase exhibits partial coherence with the metallic framework and provides a host for Li^+ ions. This spontaneously formed nanostructure offers the stability required for the cycling of micron-sized particles, obviating the necessity for the nanoscale pre-modifications usually essential for traditional metal oxides used in batteries.

Definitely, much like ionic conduction, the electronic conductivity of compositionally complex materials is influenced by a myriad of factors. These factors include the types of elements within the material and the specific characteristics of local microstates, which are determined by the extent of elemental mixing. Consequently, electrical properties can be modulated by microscopic attributes such as oxidation states, spin states, and types and concentrations of defects, as well as by external conditions like temperature and magnetic fields. In oxides where electronic conduction is thermally activated, the material's composition not only shapes its crystal and band structures but also dictates the types and densities of electronic charge carriers and their transport mechanisms. For many oxides, electron or small polaron hopping between localized states is the prevailing conduction mechanism [11]. Under these circumstances, both the compositional complexity and short-range disorder can be anticipated to affect the pre-exponential term and the activation energy for conductivity, much like they do in ionic conduction. For instance, a study by Shi et al. [12] on $AMnO_3$ perovskites found that the composition yielding the highest conductivity and lowest activation energy contained 40% acceptor constituents and displayed the smallest cation size difference. There are also in literature examples of compositionally complex oxides demonstrating band-type conduction, such as the perovskite $(Gd_{0.2}Nd_{0.2}La_{0.2}Sm_{0.2}Y_{0.2})\,CoO_3$ [13]. Its electric properties are especially influenced by the magnetic states of its A-site cations.

High-entropy materials capabilities have recently been exemplified in areas like correlated magnetism and other collective phenomena, including electronic conductivity, ferroelectricity, and phonon excitations [14-17].

In less explored areas, the intricate interplay between disordered microstates could contribute to the development of quantum materials for various applications. In fact, the inherent complexity of this promising class of materials can foster novel electronic and magnetic phases, laying the groundwork for emerging technologies like high-temperature superconductors and topological materials. However, the utilization of such 'quantum materials' necessitates an understanding of the essential elements that stabilize these unique phases.

General Thermodynamic Considerations

Designing novel High-Entropy Oxides (HEOs) generally aims to maximize the number of elements in a single-phase solid solution to obtain unique and tunable features. As discussed in this chapter, a genuine single-phase HEO becomes stable (i.e. it is an ESO) when the positive entropy of formation sufficiently offsets a disadvantageous enthalpy of formation beyond a specific temperature, resulting in an overall negative value; this is often referred as "entropy stabilization" or "entropy-driven stabilization".

In ideal mixing scenarios, the quantity of elements in a solid solution doesn't impact the solubility of any added component. However, in practical systems, factors such as the nature of extra components, structural changes, and non-ideal interactions play a role in setting the solubility boundaries. These non-ideal interactions can result in either short-range or long-range ordering, subsequently reducing the total configurational entropy of the resulting system. Given the ionic-covalent attributes of oxides, such ordering is more common than exceptional. In the few instances where ideal mixing occurs, substitutions influenced by charge coupling can alter the overall contributions to configurational entropy, especially when unique crystallographic sites are involved. Long-range ordering is minimized when oxide components with similar charges or isostructural properties are mixed. Furthermore, surface energies can also significantly impact the stability of oxide polymorphs and their solubility limits [18].

Definitely, as Nature served as the foundational expert in the synthesis of compositionally complex systems (having created through millennia complex compounds in the form of natural minerals like garnets, pyroxenes, olivines, amphiboles, under appropriate temperature and pressure conditions), thermodynamics offers several insights that can guide material scientists towards the future engineering of compositionally complex and/or high-entropy systems for practical applications:

- The simplest form of a high-entropy material consists of isostructural elements with comparable ion sizes and charges. They are stabilized exclusively by mixing entropy, counterbalancing any positive enthalpy of mixing. These HEOs maintain stability only above a particular temperature (entropy-driven stabilization) and become unstable at lower temperatures, i.e. they can be addressed as ESOs. Notable examples of ESOs include the divalent cation rocksalt system firstly discovered by Rost et al. [19], trivalent/tetravalent fluorite systems [20], and divalent/trivalent spinel-like systems [21].
- Oxide systems, due to their heightened electrostatic interactions, tend to form more stable interoxidic compounds with narrower homogeneity ranges compared to metallic systems [22].
- Given the ionic and covalent bonds present in oxides, ordering is more a rule than an exception. Both short-range and long-range ordering contribute to the reduction of overall configurational entropy. Consequently, HEOs may only be truly high-entropy systems at higher temperatures, thus generally requiring quenching as part of their fabrication processes.
- The stability of a solid solution is influenced by the types of components and their interactions rather than the sheer number of components. The term for non-ideal mixing plays a crucial role in both stabilizing and destabilizing solid solutions.
- At the nano-scale, surface energy becomes a significant contributor to Gibbs free energy, sufficiently impacting phase boundaries to stabilize specific

polymorphs and alter the scope of solid solutions. Thus, the utilization of surface energy and particle size as added variables could substantially broaden the possibilities of stabilizing certain HEO systems [22].

References

[1] Zhao, S. (2021). Role of chemical disorder and local ordering on defect evolution in high-entropy alloys. Physical Review Materials, 5(10), 103604.

[2] Jiang, B., Bridges, C.A., Unocic, R.R., Pitike, K.C., Cooper, V.R., Zhang, Y. & Page, K. (2020). Probing the local site disorder and distortion in pyrochlore high-entropy oxides. Journal of the American Chemical Society, 143(11), 4193-4204.

[3] Brahlek, M., Gazda, M., Keppens, V., Mazza, A.R., McCormack, S.J., Mielewczyk-Gryń, A. & Yamamoto, A. (2022). What is in a name: Defining "high entropy" oxides. APL Materials, 10(11).

[4] Dippo, O.F. & Vecchio, K.S. (2021). A universal configurational entropy metric for high-entropy materials. Scripta Materialia, 201, 113974.

[5] Aamlid, S.S., Oudah, M., Rottler, J. & Hallas, A.M. (2023). Understanding the role of entropy in high entropy oxides. Journal of the American Chemical Society, 145(11), 5991-6006.

[6] Yashima, M. & Takizawa, T. (2010). Atomic displacement parameters of ceria doped with rare-earth oxide $Ce_{0.8}R_{0.2}O_{1.9}$ (R = La, Nd, Sm, Gd, Y, and Yb) and correlation with oxide-ion conductivity. The Journal of Physical Chemistry C, 114(5), 2385-2392.

[7] Bérardan, D., Franger, S., Meena, A.K. & Dragoe, N. (2016). Room temperature lithium superionic conductivity in high entropy oxides. Journal of Materials Chemistry A, 4(24), 9536-9541.

[8] Wang, D., Jiang, S., Duan, C., Mao, J., Dong, Y., Dong, K., ... & Qi, X. (2020). Spinel-structured high entropy oxide $(FeCoNiCrMn)_3O_4$ as anode towards superior lithium storage performance. Journal of Alloys and Compounds, 844, 156158.

[9] Xiao, B., Wu, G., Wang, T., Wei, Z., Sui, Y., Shen, B., ... & Zheng, J. (2022). High-entropy oxides as advanced anode materials for long-life lithium-ion batteries. Nano Energy, 95, 106962.

[10] Wang, K., Hua, W., Huang, X., Stenzel, D., Wang, J., Ding, Z., ... & Mu, X. (2023). Synergy of cations in high entropy oxide lithium ion battery anode. Nature Communications, 14(1), 1487.

[11] Bosman, A.J. & Van Daal, H.J. (1970). Small-polaron versus band conduction in some transition-metal oxides. Advances in Physics, 19(77), 1-117.

[12] Shi, Y., Ni, N., Ding, Q. & Zhao, X. (2022). Tailoring high-temperature stability and electrical conductivity of high entropy lanthanum manganite for solid oxide fuel cell cathodes. Journal of Materials Chemistry A, 10(5), 2256-2270.

[13] Krawczyk, P.A., Jurczyszyn, M., Pawlak, J., Salamon, W., Baran, P., Kmita, A., ... & Żywczak, A. (2020). High-entropy perovskites as multifunctional metal oxide semiconductors: Synthesis and characterization of $(Gd_{0.2}Nd_{0.2}La_{0.2}Sm_{0.2}Y_{0.2})\,CoO_3$. ACS Applied Electronic Materials, 2(10), 3211-3220.

[14] Mazza, A.R., Skoropata, E., Sharma, Y., Lapano, J., Heitmann, T.W., Musico, B.L., ... & Ward, T.Z. (2022). Designing magnetism in high entropy oxides. Advanced Science, 9(10), 2200391.

[15] Mazza, A.R., Skoropata, E., Lapano, J., Zhang, J., Sharma, Y., Musico, B. L., ... & Ward, T.Z. (2021). Charge doping effects on magnetic properties of single-crystal La 1− x Sr x $(Cr_{0.2}Mn_{0.2}Fe_{0.2}Co_{0.2}Ni_{0.2})$ O_3 ($0\leq x\leq 0.5$) high-entropy perovskite oxides. Physical Review B, 104(9), 094204.

[16] Mazza, A.R., Gao, X., Rossi, D.J., Musico, B.L., Valentine, T.W., Kennedy, Z., ... & Ward, T.Z. (2022). Searching for superconductivity in high entropy oxide Ruddlesden–Popper cuprate films. Journal of Vacuum Science & Technology A, 40(1).

[17] Ahn, M., Park, Y., Lee, S.H., Chae, S., Lee, J., Heron, J.T., ... & Phillips, J.D. (2021). Memristors based on (Zr, Hf, Nb, Ta, Mo, W) high-entropy oxides. Advanced Electronic Materials, 7(5), 2001258.

[18] Ouyang, G., Tan, X., Wang, C.X. & Yang, G.W. (2006). Solid solubility limit in alloying nanoparticles. Nanotechnology, 17(16), 4257.

[19] Rost, C.M., Sachet, E., Borman, T., Moballegh, A., Dickey, E.C., Hou, D., ... & Maria, J.P. (2015). Entropy-stabilized oxides. Nature Communications, 6(1), 8485.

[20] Gild, J., Samiee, M., Braun, J.L., Harrington, T., Vega, H., Hopkins, P.E., ... & Luo, J. (2018). High-entropy fluorite oxides. Journal of the European Ceramic Society, 38(10), 3578-3584.

[21] Dąbrowa, J., Stygar, M., Mikuła, A., Knapik, A., Mroczka, K., Tejchman, W., ... & Martin, M. (2018). Synthesis and microstructure of the $(Co, Cr, Fe, Mn, Ni)_3O_4$ high entropy oxide characterized by spinel structure. Materials Letters, 216, 32-36.

[22] McCormack, S.J. & Navrotsky, A. (2021). Thermodynamics of high entropy oxides. Acta Materialia, 202, 1-21.

Different Classes of High-Entropy Materials

Rocksalt-Structured Oxides and High-Entropy Rocksalts

Being the second most prevalent category of materials on Earth, metal oxides find extensive applications across diverse domains. These materials serve roles ranging from cable insulators to semiconductor elements in diodes, and from constituents in glass and gel to acting as doping agents or catalysts. With advancements in nanotechnology, new attributes of metal oxides are continuously being unveiled [1]. Extensive research has been conducted on metal oxide nanoparticles, particularly because their large surface areas offer interesting possibilities for novel catalytic applications [2].

A significant body of research on rock-salt structured metal oxides, such as MgO, CaO, NiO, CoO, MnO, SrO, has been carried out on them for many different applications, ranging from catalysis (for example in the field of hydrogen storage or water splitting) [3, 4], energy storage (particularly in batteries, as electrode materials) [5, 6], electronics (for example, in magnetic recording media and spintronic devices) [7, 8], sensor technologies (particularly in gas sensors) [9], solid oxide fuel cells (particularly as electrolytes or electrode materials, due to their ionic conductivity and stability at high temperatures) [10, 11], photonics and optoelectronics (for example in applications like light-emitting diodes – LEDs- lasers, and photodetectors) [12, 13].

Magnesium oxide (MgO) serves as the archetypal metal oxide manifesting the rock-salt crystal structure, characterized by a face-centered cubic lattice arrangement, whose name derives from NaCl (see Fig. 1 for a generic rock-salt crystal structure). Each magnesium ion is surrounded by six oxygen ions, and vice versa, in an octahedral coordination geometry. MgO is characterized by a lattice constant of $a = 4.211$ Å.

Given its simple yet effective lattice structure, MgO (along with NaCl as the archetypal non oxide ionic compound) often serves as a model system for understanding the properties of more complex ionic compounds that adopt the rock-salt structure. Its wide range of applications and fundamental role in natural processes and technological applications make it a compound of significant interest in various scientific disciplines [14].

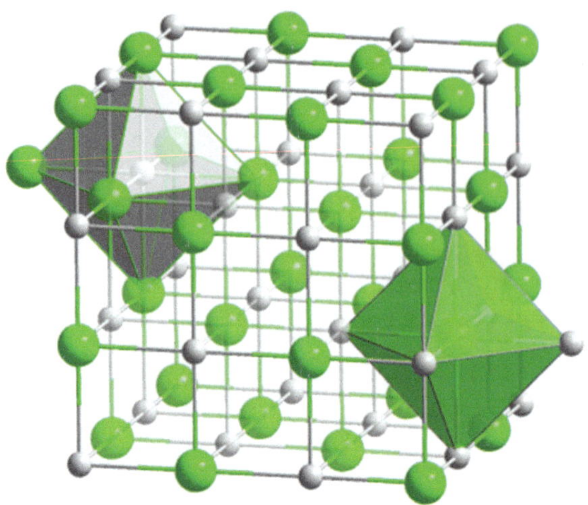

Fig. 1: A generic MO rock-salt structure (grey atoms are the metallic cations, green atoms are the oxygen anions).

Focusing on MgO, its strong catalytic and insulating properties have been the subject of numerous studies [15, 16]. It is generally found that its (100) facet is the most stable, although MgO(111) does occur naturally in the form of the mineral periclase [17]. This has sparked interest in its ion surface arrangement, and much of the experimental work has involved thin films, either as a substrate or as a coating, usually developed under high or ultra-high vacuum conditions [18, 19]. Anyway, MgO is studied both in its bulk form and as a thin film, and is intriguing for its high heat of oxide formation too, making it an ideal substrate for reactive metals [20]. Predicted to exhibit minimal electronic interactions with most metallic over-layers, MgO also serves as an excellent foundation for subsequent development of reduced-dimension metal nanostructures, whether they be 2D, 1D, or 0D [21].

Focusing on calcium oxide (CaO), one of its primary characteristics is its high thermal stability. This stability makes it an excellent choice for high-temperature applications. For instance, in metallurgical processes, CaO is often used as a flux to remove impurities such as sulfur and phosphorus from steel and other metal ores [22, 23]. By facilitating the formation of slags, it helps in purifying the metals while also protecting the furnace lining from corrosion and wear. Moreover, CaO demonstrates significant potential in desulfurization of power plants, due to its strong basicity and high reactivity towards acidic gases [24], and in the field of catalysis (as both a suitable catalyst or catalyst support for various chemical reactions, including biodiesel production) [25, 26]. Finally, a notable application of rocksalt-structured CaO is in the construction industry, where it serves as a key component in cement and concrete [27].

Another very important rock-salt structured transition metal oxide is nickel oxide (NiO), exhibiting distinct electronic properties. Its redox state can be electrochemically modulated, and it exhibits electrochromism—changing its optical properties when an electric field is applied, making it valuable for solar energy applications [28]. Additionally, nickel oxide shows promise in the fields of magnetoresistive sensors [29], as well as in catalytic processes [30, 31].

Focusing on cobalt oxide (CoO), one of its most significant application is in the field of catalysis. Its ability to exist in multiple oxidation states makes it an effective catalyst in chemical reactions, including those involved in environmental remediation and energy conversion processes [32-34]. Moreover, CoO is used as an electrode material in lithium-ion batteries, allowing for efficient intercalation and deintercalation of lithium ions, which is fundamental for the charging and discharging process of these batteries [35]. Additionally, CoO is also explored for its potential in gas sensing technologies, as its sensitivity to changes in the environment, particularly to gases like CO and H_2S, makes it a valuable material for developing sensors that can detect these gases at low concentrations [36, 37].

Rocksalt-structured strontium oxide (SrO) is another intriguing material in the realm of metal oxides. One of the primary uses of SrO is in the production of special glass and ceramics. Its high refractive index and ability to block ultraviolet (UV) rays make it a valuable component in the glass used for television and computer screens, as well as in other optical applications [38,39]. In the field of electronics, SrO is sometimes used as a component in the fabrication of thin films and substrates for superconductors [40]. Finally, in the field of catalysis, SrO can be employed as a basic catalyst for various reactions, including biodiesel production and other processes where a strong base is required [41, 42].

Definitely, rock-salt structured oxides are versatile materials that find applications in a broad spectrum of technological areas, such as catalysis, magnetoelectronic components, and gas detection systems. Their inherent characteristics, such as thermal stability, electrical conductivity, catalytic activity, and optical properties, derive directly from their crystal structure.

Furthermore, the adaptability of these materials, evident in their suitability for various modifications and enhancements, makes them a continued subject of research and innovation. As new challenges arise in technology and industry, rocksalt-structured metal oxides represent a fertile ground for future scientific exploration and technological advancement.

High-Entropy Rocksalts

Over the last two decades, cathode materials featuring the rock-salt structure have been at the forefront of academic research [43-45]. While such early research predominantly focused on chemical formulations involving two or three cations, the pioneering study by Rost et al. [46] has firstly demonstrated the possibility of stabilizing in a single-phase rocksalt structure more complex arrangements of cations through entropy, namely defining the new class of materials called High-Entropy Oxides (HEOs). Thus, the first HEO, designed by Rost et al., was an equimolar blend of MgO, CoO, NiO, CuO, and ZnO exhibiting a single-phase rock-salt crystal structure. This specific composition was chosen to offer a broad range of structural, coordinative, and cationic radius diversity to directly validate the "entropic stabilization" hypothesis. The criteria for Rost's selection included non-uniform crystal structures, electronegativities, and cation coordinations across the ensemble of binary oxides, as well as the existence of specific pairs like MgO–ZnO and CuO–NiO, which show limited mutual solubility. Moreover, all the cations in the ensemble should be isovalent, enabling continuous variation of cation ratios while maintaining overall electroneutrality at a uniform cation-to-anion ratio.

In the early exploration of high-entropy oxides, many authors focused on rock-salt structured HEOs, similarly to the one designed by Rost et al. [46]. In particular, Sarkar et al. [47] employed three distinct synthetic approaches—nebulized spray pyrolysis, flame spray pyrolysis, and reverse co-precipitation—to synthesize rock-salt structured $(Co_{0.25}Mg_{0.25}Ni_{0.25}Zn_{0.25})O$ and $(Co_{0.2}Cu_{0.2}Mg_{0.2}Ni_{0.2}Zn_{0.2})O$ HEOs. They confirmed that the structural integrity of these materials was independent of the synthesis techniques, indirectly confirming that the single-phase stabilization was dictated by the systems' entropy.

Successively, Berardan et al. [48] delved into the influence of synthetic conditions, stoichiometry, and thermal treatment on the rock-salt-structured MgCoNiCuZnO-based HEOs. They underscored the structural significance of copper, revealing that increasing Cu levels led to deviations in the Bragg peak attributes from their idealized rock-salt parameters. Such variations were ascribed to the alterations in the local coordination environment of Cu^{2+} ions.

Dupuy et al. [49] examined the effects of grain dimensions on the phase conversion of $(Co_{0.2}Cu_{0.2}Mg_{0.2}Ni_{0.2}Zn_{0.2})O$ with a rock-salt structure, observing that reduced grain sizes favored the presence of a secondary, copper-enriched tenorite phase while concurrently diminishing the temperature threshold for entropy-driven phase transformation (single-phase to multiple phases and *vice versa*).

Chen et al. [50] studied the stability of the rock-salt structured $MgCoNiCuZnO_5$ under the influence of Li and Mn doping at elevated temperatures and pressures. Their research revealed an increased compressibility of the variously-doped HEOs, which was attributed to the introduction of weaker Li-O and Mn-O bonds. This study underlined the pivotal role of dopants in modulating the mechanical and ionic characteristics of rock-salt structured HEOs, as later on Cardoso et al. [51] demonstrated that doping with Gd effectively modulated the lattice dimensions and asymmetry in the rock-salt structured HEOs.

Summarizing, HEOs featuring the rock-salt structure predominantly consist of elements such as Mg, Co, Ni, Cu, and Zn. Several fabrication techniques have been reported for these HEOs, including but not limited to, solid-state reactions [52, 53], spray pyrolysis [54], flash-aided synthesis [55, 56], microwave-assisted synthesis [57], co-precipitation [58-60] and co-precipitation coupled with hydrothermal methods [61], solution-based combustion [62, 63], and pulsed-laser deposition

[64]. These rock-salt structured HEOs are intriguing for a multitude of reasons, ranging from lithium ion conductivity and unique structural characteristics to catalytic, magnetic, electrical, thermal, and mechanical properties [65]. Finally, exhibiting elevated dielectric constants and remarkable ionic conductivity [66-68], rock-salt structured HEOs hold promise for future uses as high-k dielectric materials [69] and as solid electrolytes in battery technology [70].

High-Entropy Rocksalts as Li-Ion Batteries Electrodes

In the evolving landscape of energy storage technologies, high-entropy oxides have emerged as a promising class of materials, particularly in the context of superionic lithium conduction [68, 71-73]. Particularly, the rocksalt-structured HEOs (primarily derived by the Rost's HEO)represent a unique playground for tailoring ionic transport properties in next-generation Lithium-Ion Batteries (LIBs).

Since the first observations of lithium superionic conduction by Berardan et al. [72], lithium incorporation within the Rost HEO [46] has been deeply investigated by Spiridigliozzi et al. [74]. Particularly, lithium integrates into the high entropy rock salt structure by occupying substitutional sites, with the charge imbalance being offset by the emergence of an equivalent count of M^{3+} cations, primarily Co^{3+} and also Ni^{3+} at elevated Li-doping concentrations. Thus, the formation of the entropy-stabilized single-phase occurs without invoking the thermal characteristics usually linked to Co^{3+} reduction in the context of lithium doping. Conversely, these features become evident when lithium is absent [74]. At elevated temperatures (ranging from 1050 to 1150 °C), there is a shift in defect chemistry, resulting primarily from the conversion of Cu^{2+} to Cu^+, leading to the appearance of oxygen vacancies. Support for this conclusion is drawn from both thermogravimetric and dilatometric analyses, as well as from XPS characterizations. Moreover, within the HEO lattice, lithium resides in two distinct chemical surroundings. The first is inherently diamagnetic and essentially mirrors what is observed in conventional $LiCoO_2$ [75], while the second entails a first coordination shell composed of five to six paramagnetic neighbors. The relative intensity of the diamagnetic component amplifies with increasing lithium concentration in the HEO framework, which can be attributed to the transformation of paramagnetic Co^{2+} into diamagnetic Co^{3+}. The coexistence of these two types of environments cannot be accounted for by a random cationic distribution in the HEO lattice, hinting at some form of short-range order, possibly involving Li^+-Co^{3+} pairs.

Traditionally, graphite has served as the anode material in lithium-ion batteries due to its stable intercalation mechanism. However, its inherent capacity limitation stands at 372 mAh/g [76]. High-Entropy Oxides (HEOs) have emerged as alternative anode materials, attributed to their lithium-ion hosting capabilities through conversion-type reactions. While these materials exhibit superior capacities compared to graphite, they are marred by subpar cycle stability. Initial cycle Coulombic Efficiency (CE) often falls below 90% and can dip to as low as 50% in specific instances, as reported by Wang et al. [58]. Numerous HEOs demonstrate the unique attribute of capacity augmentation upon extended cycling, a phenomenon also observed in conventional conversion-type materials like CoO and $ZnMn_2O_4$ [77, 78].

Several theories propose that this increased capacity may result from the electrolyte's progressive infiltration into the inner oxide pores and accompanying structural alterations during cycling, as stated by Sarkar et al. [79], among others. In particular, they synthesized $(Co_{0.2}Cu_{0.2}Mg_{0.2}Ni_{0.2}Zn_{0.2})O$ using Nebulized Spray Pyrolysis (NSP) and recorded a capacity of approximately 500 mAh/g at a 0.1 A/g current. This anode material maintained a capacity of 300 mAh/g after 50 cycles and demonstrated performance in full-cell configurations with conventional cathodes [80].

Successively, Lökçü et al. [81] probed the lithium content variation in rock-salt structured $(MgCoNiZn)_{1-x}Li_xO$, that increased lithium concentration leads to the oxidation of Co^{2+} to Co^{3+} and increases the concentration of oxygen vacancies. The $(MgCoNiZn)_{0.65}Li_{0.35}O$ system, in particular, resulted in a surge in discharge capacity, boasting an initial capacity of 1930 mAh/g.

Incorporating fluorine into high-entropy rock-salts has been demonstrated to elevate their operational potential to 3.4 V [58] compared to a baseline high-entropy oxide (HEO) without fluorine, which operates at approximately 1.0 V. These fluorinated HEOs can find applications as cathode materials in lithium-ion batteries, as they exhibited an initial charge capacity of 161 mAh/g at C/10 for the $Li_x(Co_{0.2}Cu_{0.2}Mg_{0.2}Ni_{0.2}Zn_{0.2})OF_x$ system. Remarkably, this capacity exceeded that of a non-entropy stabilized $LiNiOF_x$ baseline system [58], even though it contained a higher proportion of redox-active nickel. Nonetheless, this material exhibited substantial irreversible capacity loss in the initial cycle and a low overall Coulombic Efficiency (CE). Wang et al. attributed this performance degradation to factors such as electrolyte decomposition, Solid Electrolyte Interphase (SEI) formation, and side reactions.

Subsequently, the same research group explored different layered HEOs, including $LiNa(Ni_{0.2}Co_{0.2}Mn_{0.2}Al_{0.2}Fe_{0.2})O_2$, reporting an oxidation peak at 4.0 V for it, surpassing the typical 3.8 V observed in standard $Li(NiCoMn)1O_2$ [80]. Nevertheless, these electrodes were characterized by low capacity and quick capacity degradation, dropping below 70 mAh/g after just 20 cycles. A minor addition of sodium ions aimed to improve battery cycling by enlarging the diffusion pathways. Even a 1 mol% addition of sodium has been reported to slightly enhance capacity retention in traditional layered cathodes [82].

The unconventional electrochemical behavior of those high-entropy materials has been credited to both high-entropy stabilization and the yet-unexplained phenomenon often referred to as the cocktail effect, firstly introduced for High-Entropy Alloys (HEAs) [83,84]. However, the configurational entropy inherent to HEOs, which are thermodynamically metastable at ambient temperatures, is not adequate to facilitate structural reversibility during conversion-type electrochemical reactions. To this regard, a very recent study by Wang et al. [85] shed light on the synergistic effect of multiple cations in the Rost's HEO composition (i.e. $Mg_{0.2}Co_{0.2}Ni_{0.2}Cu_{0.2}Zn_{0.2}O$) at both the atomic and nanoscale levels during electrochemical processes. This study offers insights into the so-called *cocktail effect*, revealing that elements with higher electronegativity create a chemically inert 3D metallic nano-network, thereby facilitating electron transport. The electrochemically inactive cations serve to stabilize an oxide nanophase that is semi-coherent with the metallic phase, providing a scaffold for lithium-ion accommodation. This spontaneously assembled nano-architecture allows for stable electrochemical cycling in micron-sized particles, obviating the necessity for nanoscale additional pre-treatments. Thus, it is suggested that compositional diversity at the elemental level is instrumental in optimizing multi-cation electrode materials for lithium-ion batteries.

In other words, the original HEO structure is not recoverable after the initial charging cycle; thus, the remarkable electrochemical performance observed by many researchers in rock-salt structured HEOs is derived from cationic synergy, serving three distinct functions:

- Zinc and Cobalt are the electrochemically active components, chiefly responsible for the battery's initial lithiation capacity.
- Magnesium ions are electrochemically inert and are not reduced by lithium. They serve to stabilize the oxide phase, which houses lithium ions in the discharged state and holds Cobalt and Zinc ions in the charged state. This aligns with previous studies like those by Qiu et al. [86], suggesting that inactive MgO helps preserve the electrode material.
- Copper and Nickel do not engage in redox reactions post the first discharge but sustain the nanoscale 3D metallic network, thereby maintaining excellent electronic conductivity.

This results in a significant fraction of lithium ions remaining in the oxide nanophase within the HEO system after charging, ensuring effective ionic conductivity. Ultimately, the material transitions into a composite featuring semi-coherent metal/oxide nanophases, reducing the reaction pathway to the nanometer scale and thereby facilitating rapid reactions. This also minimizes the formation of bulk Li_2O, a contributing factor to capacity loss in other battery systems [87].

Conventional multi-cation oxides composed of binary metal elements typically offer either a metallic phase in the charged state to boost electrical conductivity or dual oxide compounds for volume buffering. In contrast, multicationic HEOs feature a diverse array of cationic properties, resulting in semi-coherently intertwined metal and oxide nanophases, offering distinct advantages over traditional binary cation oxides [85]. While the post-first-discharge reactions are independent of the entropy-driven stabilization, the inherently disordered chemical environment of HEO systems is vital for their possible use as battery electrodes.

High-Entropy Rocksalts as Na-Ion Batteries Electrodes

Sodium-ion batteries (NIBs) have garnered considerable research interest, primarily due to their affordability and the abundance of raw materials, making them suitable for large-scale energy storage systems [88-90]. Over the last ten years, a variety of cathode materials suitable for NIBs have been explored, including simple oxides [91], polyanionic compounds [92], Prussian blue analogs [93] and organic compounds [94, 95]. Among these, layered oxide materials stand out as particularly promising for large-scale practical applications, thanks to their high structural adaptability for the reversible insertion and removal of Na^+ ions [96]. To date, several notable materials demonstrating diverse electrochemical performances have been developed, including Mn-based (Na_xMnO_2 [97], $Na_{0.67}Fe_{0.5}Mn_{0.5}O_2$ [98]), Ni-rich ($Na[NiFeMn]O_2$ and $Na[NiCoMn]O_2$ [99]), and Cu-based ($CuSO_4$ [100]) oxides.

High-entropy oxides (HEOs) represent a novel approach in the field of NIBs, offering advanced systems with unique characteristics unattainable with conventional materials that have one or a few dominant elements. In this context, Zhao et al. [101] recently proposed a novel HEO cathode, $Na(Ni_{0.12}Cu_{0.12}Mg_{0.12}Fe_{0.15}Co_{0.15}Mn_{0.1}Ti_{0.1}Sn_{0.1}Sb_{0.04})O_2$, featuring nine different transition metal ions ranging from divalent to pentavalent oxidation states. This high-entropy cathode provided exceptional cycling stability and significantly improved sodium storage capacity. It exhibited superior capacity retention across various current rates, maintaining about 83% capacity after 500 cycles. Such enhanced performance is likely due to the multiple elements in the high-entropy matrix, which can better accommodate local interaction changes during the sodium ion insertion and removal process.

Furthermore, Wang et al. [102] successfully synthesized three distinct layered P2-type high-entropy oxides through a solid-state reaction process for use as cathode materials in sodium-ion batteries. Their study revealed that the $Na_{0.67}(Mn_{0.45}Ni_{0.18}Co_{0.18}Ti_{0.1}Mg_{0.03}Al_{0.04}Fe_{0.02})O_2$ system displayed remarkable reversibility in sodiation/desodiation processes within the voltage ranges of 2.6–4.6 V and 1.5–4.6 V. In comparison, the low-entropy $Na_{0.67}(Mn_{0.55}Ni_{0.21}Co_{0.24})O_2$ system suffered from rapid capacity loss, while the medium-entropy $Na_{0.67}(Mn_{0.45}Ni_{0.18}Co_{0.24}Ti_{0.1}Mg_{0.03})O_2$ system experienced a more moderate decline in capacity. The increased configurational entropy in these materials helps counteract the undesired P2 $\rightarrow$ O2 phase transition for layered systems, thereby enhancing structural stability.

The findings of these first studies focusing on the rocksalt-derived HEOs as promising SIBs electrodes provide strong and promising evidence supporting their great potential in electrochemical energy storage applications.

High-Entropy Rocksalts for Enhanced Catalysis

Given the inherent surface intricacies of High-Entropy Oxides (HEOs), the rock-salt structured ones, mainly containing transition metals (like Iron, Cobalt, Nickel, Copper, and Zinc), have been studied also for their potential application in electrocatalysis [103-105]. The unique active binding sites exposed in rocksalt-structured HEOs not only facilitate a nearly seamless adsorption energy spectrum but also enable the simultaneous tuning of multiple properties.

One prevalent line of reasoning for this preference is that the 3d electronic structure of transition metals can be modulated by fabricating multi-metal oxides, which in turn enhances electrocatalytic performance. Another crucial aspect that influences electrocatalysts is electrical conductivity, as it dictates the charge transfer rate during catalytic reactions.

Therefore, the engineering of advanced electrocatalysts employing rocksalt-structured high-entropy oxides should consequently lead to improvements in both electrical conductivity and electrochemical activity.

To this regard, Gu et al. [103] proposed an innovative approach involving lithium doping for the development of Lithium Transition Metal Oxides (LTM) consisting of six evenly distributed metal elements (i.e. Li, Fe, Co, Ni, Cu, and Zn) in a rocksalt-like crystalline structure. The inclusion of lithium led to an increase in oxygen vacancies, which in turn caused a contraction of the lattice constant. Particularly, when the molar proportion of lithium matched that of the other transition metal cations, Gu et al. observed a tenfold increase in electrical conductivity. Additionally, their proposed HEO recorded optimal Hydrogen Evolution Reaction (HER) and Oxygen Evolution Reaction (OER) performances, registering 207 mV at 10 mA cm^{-2} for HER and 347 mV at 10 mA cm^{-2} for OER, bolstered by the improved electrical conductivity.

Liu et al. [104] analyzed a Rost's HEO (i.e. $Mg_{0.2}Co_{0.2}Ni_{0.2}Cu_{0.2}Zn_{0.2}O$) as an electrocatalyst for the Oxygen Evolution Reaction (OER). The resulting HEO sample showcased exceptional intrinsic OER capabilities, achieving Turnover Frequencies (TOFs) that are 15 and 84 times higher than those of CoO and NiO at a voltage of 1.65 V, respectively.

In the work of Wang et al. [106], a high-efficiency CO oxidation catalyst, featuring a distinctive heterostructure composed of a copper cerium oxide supported by the rocksalt-structured $Mg_{0.2}Co_{0.2}Ni_{0.2}Cu_{0.2}Zn_{0.2}O$ (named $CuCeO_x$-HEO) was successfully synthesized and characterized. The $CuCeO_x$-HEO exceptional thermal resilience in CO oxidation is attributable to the entropy-stabilized framework and the catalytic role played by the $CuCeO_x$ subsystem. In a comparison with a CeO_2-CuO sample that lacks HEO support, it was found by the authors that although the sample initially exhibits comparable CO oxidation activity to $CuCeO_x$-HEO, it loses this functionality at elevated temperatures due to the agglomeration of large CuO particles. Contrarily, the $CuCeO_x$-HEO heterostructure maintains its catalytic activity under ongoing high-temperature conditions. This unique stabilization mechanism may offer valuable insights for the future development of other high-temperature stable catalysts.

References

[1] Ogale, S.B., Venkatesan, T.V. & Blamire, M. (Eds.). (2013). Functional Metal Oxides: New Science and Novel Applications. John Wiley & Sons.

[2] Khan, M.M., Adil, S.F. & Al-Mayouf, A. (2015). Metal oxides as photocatalysts. Journal of Saudi Chemical Society, 19(5), 462-464.

[3] Kim, T.W., Kim, M., Kim, S.K., Choi, Y.N., Jung, M., Oh, H. & Suh, Y.W. (2021). Remarkably fast low-temperature hydrogen storage into aromatic benzyltoluenes over MgO-supported Ru nanoparticles with homolytic and heterolytic H2 adsorption. Applied Catalysis B: Environmental, 286, 119889.

[4] Li, Y., Peng, Y.K., Hu, L., Zheng, J., Prabhakaran, D., Wu, S., ... & Tsang, S.C.E. (2019). Photocatalytic water splitting by N-TiO2 on MgO (111) with exceptional quantum efficiencies at elevated temperatures. Nature Communications, 10(1), 4421.

[5] Iftikhar, M., Latif, S., Jevtovic, V., Ashraf, I.M., El-Zahhar, A.A., Saleh, E.A.M. & Abbas, S.M. (2022). Current advances and prospects in NiO-based lithium-ion battery anodes. Sustainable Energy Technologies and Assessments, 53, 102376.

[6] Cao, K., Jiao, L., Liu, Y., Liu, H., Wang, Y. & Yuan, H. (2015). Ultra-high capacity lithium-ion batteries with hierarchical CoO nanowire clusters as binder free electrodes. Advanced Functional Materials, 25(7), 1082-1089.

[7] Xia, W.X., Tohara, K., Murakami, Y., Shindo, D., Ito, T., Iwasaki, Y. & Tachibana, J. (2006). Observation of magnetization of obliquely evaporated Co-CoO magnetic recording tape. IEEE Transactions on Magnetics, 42(10), 3252-3254.

[8] C Prestgard, M., P Siegel, G. & Tiwari, A. (2014). Oxides for spintronics: A review of engineered materials for spin injection. Advanced Materials Letters, 5(5), 242-247.

[9] Dey, A. (2018). Semiconductor metal oxide gas sensors: A review. Materials Science and Engineering: B, 229, 206-217.

[10] Istomin, S.Y., Lyskov, N.V., Mazo, G.N. & Antipov, E.V. (2021). Electrode materials based on complex d-metal oxides for symmetrical solid oxide fuel cells. Russian Chemical Reviews, 90(6), 644.

[11] Zheng, K., Świerczek, K., Zając, W. & Klimkowicz, A. (2014). Rock salt ordered-type double perovskite anode materials for solid oxide fuel cells. Solid State Ionics, 257, 9-16.

[12] Yu, X., Marks, T.J. & Facchetti, A. (2016). Metal oxides for optoelectronic applications. Nature Materials, 15(4), 383-396.

[13] Gaire, M., Subedi, B., Adireddy, S. & Chrisey, D. (2020). Ultra-long cycle life and binder-free manganese-cobalt oxide supercapacitor electrodes through photonic nanostructuring. RSC Advances, 10(66), 40234-40243.

[14] Fernandes, M., RB Singh, K., Sarkar, T., Singh, P. & Pratap Singh, R. (2020). Recent applications of magnesium oxide (MgO) nanoparticles in various domains. Advanced Materials Letters, 11(8), 1-10.

[15] Saied, E., Eid, A.M., Hassan, S.E.D., Salem, S.S., Radwan, A.A., Halawa, M., ... & Fouda, A. (2021). The catalytic activity of biosynthesized magnesium oxide nanoparticles (MgO-NPs) for inhibiting the growth of pathogenic microbes, tanning effluent treatment, and chromium ion removal. Catalysts, 11(7), 821.

[16] Liu, H., Mao, X., Cui, J., Jiang, S. & Zhang, W. (2019). Investigation of high temperature electrical insulation property of MgO ceramic films and the influence of annealing process. Ceramics International, 45(18), 24343-24347.

[17] Luong, N.T., Holmboe, M. & Boily, J.F. (2023). MgO nanocube hydroxylation by nanometric water films. Nanoscale, 15(24), 10286-10294.

[18] Kerrigan, A., Pande, K., Pingstone, D., Cavill, S.A., Gajdardziska-Josifovska, M., McKenna, K.P., ... & Lazarov, V.K. (2022). Nano-faceted stabilization of polar-oxide thin films: The case of MgO (111) and NiO (111) surfaces. Applied Surface Science, 596, 153490.

[19] Vardhan, H., Sharma, G., Sharma, K., Choudhary, R.J., Phase, D.M., Gupta, M., ... & Gupta, A. (2022, February). Evolution of magnetic and structural properties of MgO thin film with vacuum annealing. *In:* IOP Conference Series: Materials Science and Engineering (Vol. 1225, No. 1, 012064). IOP Publishing.

[20] Magkoev, T.T. (2021). Formation and modification of metal oxide substrates for controlled molecular adsorption and transformation on their surface. Russian Journal of Physical Chemistry A, 95(6), 1081-1092.

[21] Huma, T., Hakimi, N., Younis, M., Huma, T., Ge, Z. & Feng, J. (2022). MgO heterostructures: From synthesis to applications. Nanomaterials, 12(15), 2668.

[22] Horie, T., Kono, T., Kishimoto, Y., Kawagishi, K., Suzuki, S. & Harada, H. (2021). The Effect of CaO-MgO mixture on desulfurization of molten Ni-base superalloy. Metallurgical and Materials Transactions B, 52(4), 2687-2702.

[23] Hashim, Z.H., Kuwahara, Y., Hanaki, A., Mohamed, A.R. & Yamashita, H. (2023). Synthesis of a $CaO-Fe_2O_3-SiO_2$ composite from a dephosphorization slag for adsorption of CO_2. Catalysis Today, 410, 264-272.

[24] Li, X., Chen, J., Zhao, R., Xiong, Z. & Lu, C. (2022). Determination on the activity of formed $CaSO_4$ for arsenic adsorption during arsenic capture by CaO with the presence of SO_2: Experimental and density functional theory study. Chemical Engineering Journal, 427, 132015.

[25] Das, V., Tripathi, A.M., Borah, M.J., Dunford, N.T. & Deka, D. (2020). Cobalt-doped CaO catalyst synthesized and applied for algal biodiesel production. Renewable Energy, 161, 1110-1119.

[26] Huo, L., Wang, T., Pu, Y., Li, C., Li, L., Zhai, M., ... & Bai, Y. (2021). Effect of cobalt doping on the stability of CaO-based catalysts for dimethyl carbonate synthesis via the transesterification of propylene carbonate with methanol. ChemistrySelect, 6(38), 10226-10237.

[27] Hewlett, P. & Liska, M. (Eds.). (2019). Lea's Chemistry of Cement and Concrete. Butterworth-Heinemann.

[28] Wang, J., Huo, X., Guo, M. & Zhang, M. (2022). A review of NiO-based electrochromic-energy storage bifunctional material and integrated device. Journal of Energy Storage, 47, 103597.

[29] Chatterjee, S., Maiti, R. & Chakravorty, D. (2020). Large magnetodielectric effect and negative magnetoresistance in NiO nanoparticles at room temperature. RSC Advances, 10(23), 13708-13716.

[30] Alnarabiji, M.S., Tantawi, O., Ramli, A., Zabidi, N.A.M., Ghanem, O.B. & Abdullah, B. (2019). Comprehensive review of structured binary Ni-NiO catalyst:

Synthesis, characterization and applications. Renewable and Sustainable Energy Reviews, 114, 109326.

[31] Dey, S. & Mehta, N.S. (2020). Oxidation of carbon monoxide over various nickel oxide catalysts in different conditions: A review. Chemical Engineering Journal Advances, 1, 100008.

[32] Zhou, F., Liu, M., Li, X., Zhu, D., Ma, Y., Qu, X., ... & Yin, H. (2023). Preparation of CoO-C catalysts from spent lithium-ion batteries and waste biomass for efficient degradation of ciprofloxacin via peroxymonosulfate activation. Chemical Engineering Journal, 471, 144469.

[33] Xu, Y., Long, J., Tu, L., Dai, W., Yang, L., Zou, J., ... & Luo, S. (2021). CoO engineered Co9S8 catalyst for CO_2 photoreduction with accelerated electron transfer endowed by the built-in electric field. Chemical Engineering Journal, 426, 131849.

[34] Have, I.C.T., Kromwijk, J.J., Monai, M., Ferri, D., Sterk, E.B., Meirer, F. & Weckhuysen, B.M. (2022). Uncovering the reaction mechanism behind CoO as active phase for CO_2 hydrogenation. Nature Communications, 13(1), 324.

[35] Li, H., Hu, Z., Xia, Q., Zhang, H., Li, Z., Wang, H., ... & Miao, G.X. (2021). Operando magnetometry probing the charge storage mechanism of CoO lithium-ion batteries. Advanced Materials, 33(12), 2006629.

[36] Vioto, G.C., Perfecto, T.M., Zito, C.A. & Volanti, D.P. (2020). Enhancement of 2-butanone sensing properties of SiO2@ CoO core-shell structures. Ceramics International, 46(14), 22692-22698.

[37] Hu, Q.M., Dong, Z., Zhang, G.X., Li, Y.X., Xing, S.F., Ma, Z.H., ... & Xu, J.Q. (2023). Ultra-thin ALD CoO x-ZnO heterogenous films as highly sensitive and environmentally friendly H_2S sensor. Rare Metals, 42(9), 3054-3063.

[38] Ismail, M.M., Abo-Mosallam, H.A. & Darwish, A.G. (2022). Influence of SrO on structural, optical and electrical properties of $LiF-MgO-Bi_2O_3-SiO_2$ glasses for energy storage applications. Journal of Non-Crystalline Solids, 590, 121667.

[39] Balasubramani, V., Pham, P.V., Ibrahim, A., Hakami, J., Ansari, M.Z. & Le, T.K. (2022). Enhanced photosensitive of Schottky diodes using SrO interfaced layer in MIS structure for optoelectronic applications. Optical Materials, 129, 112449.

[40] Li, Z., Sang, L., Liu, P., Yue, Z., Fuhrer, M.S., Xue, Q. & Wang, X. (2021). Atomically thin superconductors. Small, 17(9), 1904788.

[41] El-Sayed, F., Hussien, M.S., AlAbdulaal, T.H., Abdel-Aty, A.H., Zahran, H.Y., Yahia, I.S. & Elhaes, H. (2022). Study of catalytic activity of G-SrO nanoparticles for degradation of cationic and anionic dye and comparative study photocatalytic and electro & photo-electrocatalytic of anionic dye degradation. Journal of Materials Research and Technology, 20, 959-975.

[42] Palitsakun, S., Koonkuer, K., Topool, B., Seubsai, A. & Sudsakorn, K. (2021). Transesterification of Jatropha oil to biodiesel using SrO catalysts modified with CaO from waste eggshell. Catalysis Communications, 149, 106233.

[43] Li, L., Lun, Z., Chen, D., Yue, Y., Tong, W., Chen, G., ... & Wang, C. (2021). Fluorination-enhanced surface stability of cation-disordered rocksalt cathodes for Li-ion batteries. Advanced Functional Materials, 31(25), 2101888.

[44] Chen, D., Ahn, J. & Chen, G. (2021). An overview of cation-disordered lithium-excess rocksalt cathodes. ACS Energy Letters, 6(4), 1358-1376.

[45] Clément, R.J., Lun, Z. & Ceder, G. (2020). Cation-disordered rocksalt transition metal oxides and oxyfluorides for high energy lithium-ion cathodes. Energy & Environmental Science, 13(2), 345-373.

[46] Rost, C.M., Sachet, E., Borman, T., Moballegh, A., Dickey, E.C., Hou, D., ... & Maria, J.P. (2015). Entropy-stabilized oxides. Nature Communications, 6(1), 8485.

[47] Sarkar, A., Djenadic, R., Usharani, N.J., Sanghvi, K.P., Chakravadhanula, V.S., Gandhi, A.S., Hahn, H. & Bhattacharya, S.S. (2017). Nanocrystalline multicomponent entropy stabilised transition metal oxides. Journal of European Ceramic Society, 37, 747-754.

[48] Berardan, D., Meena, A., Franger, S., Herrero, C. & Dragoe, N.(2017). Controlled Jahn-Teller distortion in (MgCoNiCuZn) O-based high entropy oxides. Journal of Alloys and Compounds, 704, 693-700.

[49] Dupuy, A.D., Wang, X. & Schoenung, J.M. (2019). Entropic phase transformation in nanocrystalline high entropy oxides. Materials Research Letters, 7, 60-67.

[50] Chen, J., Liu, W., Liu, J., Zhang, X., Yuan, M., Zhao, Y.,... & Kunz, M. (2019). Stability and compressibility of cation-doped high-entropy oxide $MgCoNiCuZnO_5$. The Journal of Physical Chemistry C, 123, 17735-17744.

[51] Cardoso, A.L., Perdomo, C.P., Kiminami, R.H. & Gunnewiek, R.F. (2021). Enhancing the stabilization of nanostructured rocksalt-like high entropy oxide by Gd addition. Materials Letters, 285, 129175.

[52] Dupuy, A.D., Wang, X. & Schoenung, J.M. (2019). Entropic phase transformation in nanocrystalline high entropy oxides. Materials Research Letters, 7(2), 60-67.

[53] Ma, Y., Zhang, S., Yang, X. & Wang, Q. (2023). Preparation and electromagnetic absorption properties of bulk (MgCoNiCuZn)O. Journal of Electronic Materials, 52(10), 959-967.

[54] Phakatkar, A.H., Saray, M.T., Rasul, M.G., Sorokina, L.V., Ritter, T.G., Shokuhfar, T. & Shahbazian-Yassar, R. (2021). Ultrafast synthesis of high entropy oxide nanoparticles by flame spray pyrolysis. Langmuir, 37(30), 9059-9068.

[55] Kumar, A., Sharma, G., Aftab, A. & Ahmad, M.I. (2020). Flash assisted synthesis and densification of five component high entropy oxide (Mg, Co, Cu, Ni, Zn) O at 350° C in 3 min. Journal of the European Ceramic Society, 40(8), 3358-3362.

[56] Liu, D., Peng, X., Liu, J., Chen, L., Yang, Y. & An, L. (2020). Ultrafast synthesis of entropy-stabilized oxide at room temperature. Journal of the European Ceramic Society, 40(6), 2504-2508.

[57] Kheradmandfard, M., Minouei, H., Tsvetkov, N., Vayghan, A.K., Kashani-Bozorg, S.F., Kim, G., ... & Kim, D.E. (2021). Ultrafast green microwave-assisted synthesis of high-entropy oxide nanoparticles for Li-ion battery applications. Materials Chemistry and Physics, 262, 124265.

[58] Wang, Q., Sarkar, A., Wang, D., Velasco, L., Azmi, R., Bhattacharya, S.S., ... & Breitung, B. (2019). Multi-anionic and-cationic compounds: New high entropy materials for advanced Li-ion batteries. Energy & Environmental Science, 12(8), 2433-2442.

[59] Yv, L., Wang, J., Shi, Z., Shi, J. & Wang, X. (2022). Preparation of high-entropy ceramic oxide powder at different calcination temperatures and electrochemical applications. Ceramics International, 48(18), 26370-26377.

[60] Spiridigliozzi, L., Ferone, C., Cioffi, R. & Dell'Agli, G. (2021). A simple and effective predictor to design novel fluorite-structured High Entropy Oxides (HEOs). Acta Materialia, 202, 181-189.

[61] Biesuz, M., Spiridigliozzi, L., Dell'Agli, G., Bortolotti, M. & Sglavo, V.M. (2018). Synthesis and sintering of (Mg, Co, Ni, Cu, Zn) O entropy-stabilized oxides obtained by wet chemical methods. Journal of Materials Science, 53(11), 8074-8085.

[62] Mao, A., Xiang, H.Z., Zhang, Z.G., Kuramoto, K., Yu, H. & Ran, S. (2019). Solution combustion synthesis and magnetic property of rock-salt ($Co_{0.2}Cu_{0.2}Mg_{0.2}Ni_{0.2}Zn_{0.2}$) O high-entropy oxide nanocrystalline powder. Journal of Magnetism and Magnetic Materials, 484, 245-252.

[63] Aydinyan, S., Kirakosyan, H., Sargsyan, A., Volobujeva, O. & Kharatyan, S. (2022). Solution combustion synthesis of MnFeCoNiCu and (MnFeCoNiCu) $3O_4$ high entropy materials and sintering thereof. Ceramics International, 48(14), 20294-20305.

[64] Guo, H., Wang, X., Dupuy, A.D., Schoenung, J.M. & Bowman, W.J. (2022). Growth of nanoporous high-entropy oxide thin films by pulsed laser deposition. Journal of Materials Research, 1-12.

[65] Akrami, S., Edalati, P., Fuji, M. & Edalati, K. (2021). High-entropy ceramics: Review of principles, production and applications. Materials Science and Engineering: R: Reports, 146, 100644.

[66] Bérardan, D., Franger, S., Dragoe, D., Meena, A.K. & Dragoe, N. (2016). Colossal dielectric constant in high entropy oxides. Physica Status Solidi – Rapid Research Letters (RRL), 10(4), 328-333.

[67] Salian, A., Praveen, L.L. & Mandal, S. (2023). Role of Mg–O on phase stabilization in solution combustion processed rocksalt structured high entropy oxide (CoCuMgZnNi) O with high dielectric performance. Ceramics International, 49(19), 31131-31143.

[68] Biesuz, M., Chen, J., Bortolotti, M., Speranza, G., Esposito, V. & Sglavo, V.M. (2022). Ni-free high-entropy rock salt oxides with Li superionic conductivity. Journal of Materials Chemistry A, 10(44), 23603-23616.

[69] Zhang, J., Yan, J., Calder, S., Zheng, Q., McGuire, M.A., Abernathy, D.L., ... & Hermann, R.P. (2019). Long-range antiferromagnetic order in a rocksalt high entropy oxide. Chemistry of Materials, 31(10), 3705-3711.

[70] Cai, Z.P., Ma, C., Kong, X.Y., Wu, X.Y., Wang, K.X. & Chen, J.S. (2022). High-performance PEO-based all-solid-state battery achieved by Li-conducting high entropy oxides. ACS Applied Materials & Interfaces, 14(51), 57047-57054.

[71] Strauss, F., Lin, J., Duffiet, M., Wang, K., Zinkevich, T., Hansen, A.L., ... & Brezesinski, T. (2022). High-entropy polyanionic lithium superionic conductors. ACS Materials Letters, 4(2), 418-423.

[72] Bérardan, D., Franger, S., Meena, A.K. & Dragoe, N. (2016). Room temperature lithium superionic conductivity in high entropy oxides. Journal of Materials Chemistry A, 4(24), 9536-9541.

[73] Osenciat, N., Bérardan, D., Dragoe, D., Léridon, B., Holé, S., Meena, A.K., ... & Dragoe, N. (2019). Charge compensation mechanisms in Li-substituted high-entropy oxides and influence on Li superionic conductivity. Journal of the American Ceramic Society, 102(10), 6156-6162.

[74] Spiridigliozzi, L., Dell'Agli, G., Callone, E., Dirè, S., Campostrini, R., Bettotti, P., ... & Biesuz, M. (2022). A structural and thermal investigation of Li-doped high entropy (Mg, Co, Ni, Cu, Zn) O obtained by co-precipitation. Journal of Alloys and Compounds, 927, 166933.

[75] Antolini, E. (2004). $LiCoO_2$: Formation, structure, lithium and oxygen nonstoichiometry, electrochemical behaviour and transport properties. Solid State Ionics, 170(3-4), 159-171.

[76] Nitta, N., Wu, F., Lee, J.T. & Yushin, G. (2015). Li-ion battery materials: Present and future. Materials Today, 18(5), 252-264.

[77] Wu, J.B., Tu, J.P., Han, T.A., Yang, Y.Z., Zhang, W.K. & Zhao, X.B. (2006). High-rate dischargeability enhancement of Ni/MH rechargeable batteries by addition of nanoscale CoO to positive electrodes. Journal of Power Sources, 156(2), 667-672.

[78] Alfaruqi, M.H., Rai, A.K., Mathew, V., Jo, J. & Kim, J. (2015). Pyro-synthesis of nanostructured spinel $ZnMn_2O_4$/C as negative electrode for rechargeable lithium-ion batteries. Electrochimica Acta, 151, 558-564.

[79] Sarkar, A., Velasco, L., Wang, D.I., Wang, Q., Talasila, G., de Biasi, L., ... & Breitung, B. (2018). High entropy oxides for reversible energy storage. Nature Communications, 9(1), 3400.

[80] Wang, Q., Sarkar, A., Li, Z., Lu, Y., Velasco, L., Bhattacharya, S.S., ... & Breitung, B. (2019). High entropy oxides as anode material for Li-ion battery applications: A practical approach. Electrochemistry Communications, 100, 121-125.

[81] Lokcu, E., Toparli, C. & Anik, M. (2020). Electrochemical performance of (MgCoNiZn) 1–x Li x O high-entropy oxides in lithium-ion batteries. ACS Applied Materials & Interfaces, 12(21), 23860-23866.

[82] He, T., Chen, L., Su, Y., Lu, Y., Bao, L., Chen, G., ... & Wu, F. (2019). The effects of alkali metal ions with different ionic radii substituting in Li sites on the electrochemical properties of Ni-Rich cathode materials. Journal of Power Sources, 441, 227195.

[83] Cao, B.X., Wang, C., Yang, T. & Liu, C.T. (2020). Cocktail effects in understanding the stability and properties of face-centered-cubic high-entropy alloys at ambient and cryogenic temperatures. Scripta Materialia, 187, 250-255.

[84] Pickering, E.J. & Jones, N.G. (2016). High-entropy alloys: A critical assessment of their founding principles and future prospects. International Materials Reviews, 61(3), 183-202.

[85] Wang, K., Hua, W., Huang, X., Stenzel, D., Wang, J., Ding, Z., ... & Mu, X. (2023). Synergy of cations in high entropy oxide lithium ion battery anode. Nature Communications, 14(1), 1487.

[86] Qiu, N., Chen, H., Yang, Z., Sun, S., Wang, Y. & Cui, Y. (2019). A high entropy oxide $(Mg_{0.2}Co_{0.2}Ni_{0.2}Cu_{0.2}Zn_{0.2}O)$ with superior lithium storage performance. Journal of Alloys and Compounds, 777, 767-774.

[87] Xu, L., Tang, S., Cheng, Y., Wang, K., Liang, J., Liu, C., ... & Mai, L. (2018). Interfaces in solid-state lithium batteries. Joule, 2(10), 1991-2015.

[88] Liu, T., Zhang, Y., Jiang, Z., Zeng, X., Ji, J., Li, Z., ... & Liang, C. (2019). Exploring competitive features of stationary sodium ion batteries for electrochemical energy storage. Energy & Environmental Science, 12(5), 1512-1533.

[89] Eftekhari, A. & Kim, D.W. (2018). Sodium-ion batteries: New opportunities beyond energy storage by lithium. Journal of Power Sources, 395, 336-348.

[90] Hirsh, H.S., Li, Y., Tan, D.H., Zhang, M., Zhao, E. & Meng, Y.S. (2020). Sodium-ion batteries paving the way for grid energy storage. Advanced Energy Materials, 10(32), 2001274.

[91] Shivaramaiah, R., Tallapragada, S., Nagabhushana, G.P. & Navrotsky, A. (2019). Synthesis and thermodynamics of transition metal oxide based sodium ion cathode materials. Journal of Solid State Chemistry, 280, 121011.

[92] Sapra, S.K., Pati, J., Dwivedi, P.K., Basu, S., Chang, J.K. & Dhaka, R.S. (2021). A comprehensive review on recent advances of polyanionic cathode materials in Na-ion batteries for cost effective energy storage applications. Wiley Interdisciplinary Reviews: Energy and Environment, 10(5), e400.

[93] Zhou, A., Cheng, W., Wang, W., Zhao, Q., Xie, J., Zhang, W., ... & Li, J. (2021). Hexacyanoferrate-type Prussian blue analogs: Principles and advances toward high-performance sodium and potassium ion batteries. Advanced Energy Materials, 11(2), 2000943.

[94] Holguin, K., Mohammadiroudbari, M., Qin, K. & Luo, C. (2021). Organic electrode materials for non-aqueous, aqueous, and all-solid-state Na-ion batteries. Journal of Materials Chemistry A, 9(35), 19083-19115.

[95] Mohammadiroudbari, M., Qin, K. & Luo, C. (2022). Multi-functionalized polymers as organic cathodes for sustainable sodium/potassium-ion batteries. Batteries & Supercaps, 5(6), e202200021.

[96] Liu, Y.F., Han, K., Peng, D.N., Kong, L.Y., Su, Y., Li, H.W., ... & Wu, X.W. (2023). Layered oxide cathodes for sodium-ion batteries: From air stability, interface chemistry to phase transition. InfoMat, e12422.

[97] TA, T., Pham, L.D., Nguyen, H.S., Hoang, C.V., LE, C., Dang, C.T., ... & Van Nguyen, N. (2017). Electrochemical performance of $Na_{0.44}MnO_2$ synthesized by hydrothermal method using as a cathode material for sodium ion batteries. Communications in Physics, 27(2), 143-149.

[98] Jiao, J., Wu, K., Li, N., Zhao, E., Yin, W., Hu, Z., ... & Xiao, X. (2022). Tuning anionic redox activity to boost high-performance sodium-storage in low-cost $Na_{0.67}Fe_{0.5}Mn_{0.5}O_2$ cathode. Journal of Energy Chemistry, 73, 214-222.

[99] Zhang, S., Li, X., Su, Y., Yang, Y., Yu, H., Wang, H., ... & Hu, Y.S. (2023). Four-in-one strategy to boost the performance of Nax [Ni, Mn] O_2. Advanced Functional Materials, 33(36), 2301568.

[100] Lee, Y., Jo, C.H., Yoo, J.K., Choi, J.U., Ko, W., Park, H., ... & Kim, J. (2020). New conversion chemistry of $CuSO_4$ as ultra-high-energy cathode material for rechargeable sodium battery. Energy Storage Materials, 24, 458-466.

[101] Zhao, C., Ding, F., Lu, Y., Chen, L. & Hu, Y.S. (2020). High-entropy layered oxide cathodes for sodium-ion batteries. Angewandte Chemie International Edition, 59(1), 264-269.

[102] Wang, J., Dreyer, S.L., Wang, K., Ding, Z., Diemant, T., Karkera, G., ... & Wang, Q. (2022). P2-type layered high-entropy oxides as sodium-ion cathode materials. Materials Futures, 1(3), 035104.

[103] Gu, Y., Bao, A., Wang, X., Chen, Y., Dong, L., Liu, X., ... & Qi, X. (2022). Engineering the oxygen vacancies of rocksalt-type high-entropy oxides for enhanced electrocatalysis. Nanoscale, 14(2), 515-524.

[104] Liu, F., Yu, M., Chen, X., Li, J., Liu, H. & Cheng, F. (2022). Defective high-entropy rocksalt oxide with enhanced metal–oxygen covalency for electrocatalytic oxygen evolution. Chinese Journal of Catalysis, 43(1), 122-129.

[105] Kim, K.H. & Choi, Y.H. (2022). Effect of constituent cations on the electrocatalytic oxygen evolution reaction in high-entropy oxide $(Mg_{0.2}Fe_{0.2}Co_{0.2}Ni_{0.2}Cu_{0.2})$ O. Journal of Electroanalytical Chemistry, 922, 116737.

[106] Wang, Y., Mi, J. & Wu, Z.S. (2022). Recent status and challenging perspective of high entropy oxides for chemical catalysis. Chem Catalysis, 2(7), 1624-1656.

Fluorite-Structured Oxides and High-Entropy Fluorites

"Fluorite" is a term that springs from the mineral CaF_2, commonly identified by the same name. A great variety of oxides adopt either the fluorite structure or structures akin to fluorite, holding a considerable and diverse technological value [1].

Fluorite-structured oxides showcase a wide range of compositions and structural variations. Specifically, they have foundational structures like the cubic fluorite (AO_2) and the disordered defect fluorite (A_4O_7). Figure 1 shows the unit cell of a generic fluorit-like crystal structure. Other derivative superstructures like the cubic pyrochlore ($A_2B_2O_7$) and the orthorhombic weberite-type (A_3BO_7) also exist, each having their unique arrangement of ordered cations and oxygen vacancies [2]. The ability of these structures to undergo phase transitions provides opportunities to modify their thermal, mechanical, and various other physical properties, even including attributes like radiation resistance [3-5].

In particular, fluorite-structured oxides (or simply fluorite oxides) are multifaceted in their application across various scientific domains like solid ionic conductors, coatings resilient to high temperatures, and catalysts [6-10]. The trio of zirconia (ZrO_2), hafnia (HfO_2), and ceria (CeO_2) constitutes the cornerstone of the most prevalent fluorite oxides. Often these are augmented with an assortment of other oxide dopants like yttria (Y_2O_3), rare earth oxides, as well as compounds like MgO, CaO, and their likes [11-15]. More generally, the MO_2 oxides of IVB group elements (excluding titanium) and actinides exhibit the crystalline fluorite-like structure.

Ionic (as they are the fluorite-structured oxides) crystal structures' stability has been expounded by Pauling, through a collection of guidelines [16], and among these, it can be stated that in stable fluorite oxides, around each cation, an arrangement of anions forms a coordinated polyhedron. The separation between the cation and anion is subject to the sum of their radii, and the ligancy of the cation is governed by their radius ratio.

A more visual grasp of this concept may be attained by envisioning the ions as spheres, with anions encircling a solitary cation. This chosen coordination of anions is influenced by the proportion of the cation's size to the anionic radius. For instance, a cation coordinated by an octahedron possesses a minimum radius that is 0.414 times the anion's radius [17]. At this ratio of distances, the anions are in touch and cannot move closer. Should the cation's size fall below this, the anion-anion distance begins to define the Coulomb energy, presuming the cation's central position remains intact. Nevertheless, opting for a structure with a lesser coordination number enables a reduced minimum anion-cation distance, enhancing the Coulomb interaction. An alternative interpretation of this instability is that when the cation is undersized for a specific coordination, it isn't "fully coordinated" but possesses space to "rattle" around [18].

The ratio between the radii of cations and anions can thus be leveraged to pinpoint the most stable coordinated polyhedra. Table 1 enumerates the critical ratio values, where the term "ligancy" is synonymous with "coordination number"—that is, the count of atoms bonded to a central atom [16]. According to this methodology, if the actual ratio of a fluorite system falls beneath any of the critical values presented in Table 1, the next lower structure turns preferable.

Table 1: Minimum ionic radius ratios for coordinated polyhedra stability in fluorite-structured systems [17]

Polyhedron	Ligancy	Minimum radius ratio	Example compound
Cubo-octahedron	12	≥ 1.000	None exists
Cube	8	≥ 0.732	CaF_2
Octahedron	6	≥ 0.414	$NaCl$
Tetrahedron	4	≥ 0.225	ZnS

As an example to that, ThO_2, an oxide crystallizing in the fluorite structure, demonstrates an eightfold coordinated ionic radius for Th^{4+} of 1.05 Å, and a fourfold coordinated radius for O^{2-} of 1.38 Å [19]. Therefore, the minimum radius ratio is 0.761, notably above the minimum required for cubic coordination, yet markedly below the critical value of 1.000 needed to fashion a cubo-octahedron structure. This ensures that Th atoms will engage in cubic coordination with O atoms. With Th atoms packed in a face-centered cubic (FCC) arrangement, a tetrahedron forms around each O atom, leading to the Fm3m space group.

Fluorite oxides are generally characterized by properties such as elevated oxygen conductivities [20, 21], diminished thermal conductivities [22, 23], remarkable hardness [24], and towering melting points [25].

Among these, the rare earth-doped zirconia and ceria are hailed for their technological prominence, owing to their heightened ionic or hybrid (ionic/electronic) conductivities coupled with their subdued thermal conductivities. They find frequent applications in devices like oxygen sensors [26, 27], electrochemical oxygen pumps [28, 29], and fuel cells [30-33]. The ionic conductivity of these fluorite oxides is strongly reliant on the nature and concentration of the dopant [20]. The maximum conductivity is generally realized by alloying with dopants

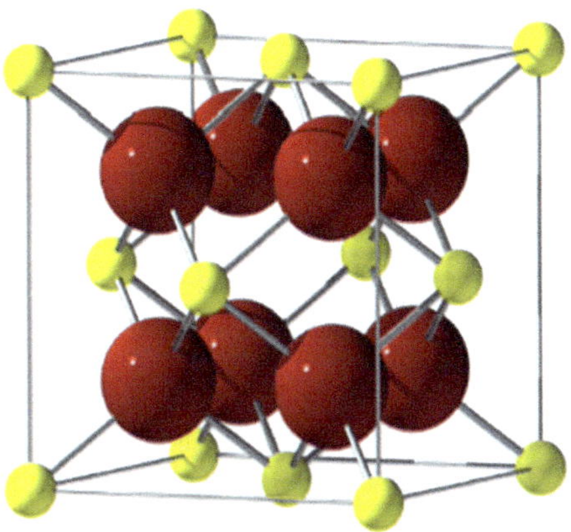

Fig. 1: Unit cell of a generic MO_2 fluorite structure (yellow atoms are the cations, red atoms are the anions)..

that induce minimal expansion or contraction within the fluorite lattice. Hence, the understanding of the influence of various dopants on the lattice parameters and their solubility limits in the cubic fluorite phase becomes essential in the study of oxygen-ion conductors.

Historical models have been utilized to predict the alterations in lattice parameters within fluorite oxide solid solutions, taking into consideration the dopants' impact on the fluorite lattice geometry. Yet, these models have been applied mainly to two fluorite oxide systems, HfO_2 [34] and ZrO_2 [34, 35], and their applicability to other fluorite oxide systems remains ambiguous. Contrasting HfO_2 and ZrO_2, other MO_2 fluorite oxides, including CeO_2, ThO_2, and UO_2, can retain the cubic fluorite structure at room temperature even without forming a solid solution.

Additionally, the precision of these models in predicting the lattice parameters is not adequately accurate for determining the dependency of ionic conductivity on dopant size within fluorite-structure oxides.

For a designated dopant valency (commonly di- or trivalent) and concentration, the oxygen-ion conductivity of fluorite oxides is influenced by the size of the dopant. It's elucidated that the effect of dopant cation size on conductivity relates to its impact on the dopant-oxygen-ion vacancy association energy, governing the activation enthalpy for conduction. Consequently, a dopant that minimizes the ionic size discrepancy between the host and dopant maximizes conductivity. However, this general assertion regarding the dopant size effect on conductivity is insufficient to elucidate conductivities in some systems, including CeO_2-M_2O and CeO_2-alkaline-earth oxides.

Similarly, the solubility limits of periodic group elements in fluorite MO_2 oxides, linked to the ionic size difference between the dopant cation and host cation, can't be rationalized merely by considering the ionic size difference. Inconsistent findings are seen in systems such as CeO_2-alkaline-earth oxide, ThO_2-lanthanide oxide (Ln_2O_3), and ZrO_2-alkaline-earth oxide [36, 37].

In light of these challenges, Kim [38] introduced empirical equations aiming to predict the changes in lattice parameters of fluorite-structure MO_2 oxide solid solutions as functions of dopant concentration. These equations are employed to shed light on the role of dopant size in dictating ionic conductivity and to clarify the relationship between the solubility limit of dopants and crystal chemical parameters in the fluorite-type MO_2 oxide solid solutions. In other words, it is an attempt to provide a more general and accurate criterion for predicting conductivities and solubility limits, surpassing previous approaches that may have lacked in comprehensiveness and accuracy.

In the fluorite structure, cations are eightfold coordinated with their neighboring anions, and each anion is tetrahedrally surrounded by four cations. The solid solution of fluorite-type MO_2 oxides is formed by substituting solute cations for host cations. When doping the fluorite oxide with aliovalent cations (those having a smaller valency than the host cation), oxygen-ion vacancies are created to maintain electrical neutrality in the altered fluorite lattice.

The changes in lattice constants due to this substitution have been the subject of research, focusing on the impact of dopant cation size on the geometry of the fluorite structure unit cell. However, the first models were limited in scope and did not accurately take into account the influence of oxygen-ion vacancies created by Y_2O_3 doping on the lattice geometry.

Glushkova et al. [34] made a more comprehensive effort by considering both the radius difference and anion vacancy formation's impact on the change of the lattice parameter. They incorporated the lattice's contraction due to oxygen-ion vacancies by considering an effective oxygen-ion radius that decreases as the dopant concentration increases. They provided an equation for the effective radius, using specific constants, including a correction factor for the association of vacancies.

This approach showed agreement with experimental values for ZrO_2- and HfO_2-Ln_2O_3 systems to within an error of 0.003 nm, but it resulted in more significant deviations in other systems. The correction factor used was not valid in ThO_2- and CeO_2-Ln_2O_3 systems, leading to greater deviations. The deviations were even more pronounced with alkaline-earth dopants due to their smaller cation valency.

Assuming that Vegard's law [39] (a rule stating that the lattice parameter of a solid solution at a given temperature is a linear function of composition) holds true, a linear relationship was assumed to exist between the solute concentration and the lattice constants. Multiple regression analyses were performed to correlate the changes in the lattice constants of fluorite-structure MO_2 solid solutions with differences in crystal chemical parameters like ionic radius, valency, and electronegativity of the dopant and host cations.

By multiplying these differences by dopant concentration and using them as independent variables, a statistical model was developed to predict the lattice constant changes [38], even though achieving an entirely accurate prediction of changes in lattice parameters of doped/co-doped fluorite oxides remains an open challenge, indicating that further refinement and consideration of additional variables may still be required.

The results of the regression analyses proposed by Kim led to a significant observation that the electronegativity parameter did not contribute to predicting the lattice parameter changes. This finding likely stems from the fact that electronegativity is related to the radius and valency of the atoms, so it doesn't provide additional independent information.

Based on these analyses, the empirical equations for the changes in the lattice parameters of fluorite oxide solid solutions were formulated. These equations are a function of the differences in ionic radius and valency of the dopant and the host cation, as well as the mole percent of the dopant in the form of MO_2:

$$d_{Hf} = 0.5098 + \sum_k (0.0203\Delta r_k + 0.00022\Delta z_k)m_k \tag{1}$$

$$d_{Zr} = 0.5120 + \sum_k (0.0212\Delta r_k + 0.00023\Delta z_k)m_k \tag{2}$$

$$d_{Ce} = 0.5413 + \sum_k (0.0220\Delta r_k + 0.00015\Delta z_k)m_k \tag{3}$$

$$d_{Th} = 0.5596 + \sum_k (0.0212\Delta r_k + 0.00011\Delta z_k)m_k \tag{4}$$

Where d (in nm) is the lattice constant of the fluorite oxide solid solution at room temperature, Δr_k (in nm) is the ionic radii difference between the kth dopant and the host cation in eightfold coordination, Δz_k is related to the valency difference between the kth dopant and the host cation in eightfold coordination, and m_k is the mole percent of the kth dopant in the form of MO_2, derived from:

$$m_k = \frac{n_k M_k}{100 + \sum_k (n_k - 1) M_k} \times 100 \tag{5}$$

Where n_k is the number of cations in the kth solute oxide and M_k the mole percent of the kth dopant oxide.

Equations (1) through (4) can be used to predict the lattice parameters of fluorite oxide solid solutions for various systems.

The difference between the observed lattice constants and the one calculated with Kim's equations tops out at 0.0014 nm.

The close match between the calculated and observed parameters makes it feasible to extrapolate the hypothetical lattice constants of pure fluorite-structured HfO_2 and ZrO_2 at room temperature using Equations (1) and (2). These formulas prove effective in determining the lattice parameters within the ternary systems of HfO_2, ZrO_2, and CeO_2.

Furthermore, Equations (1) through (4) allow us to determine the critical ionic radius of the dopant, referred to as r_{cr}, within binary systems of fluorite-structured MO_2 oxides. In particular, r_{cr} corresponds to the dopant's ionic radius, where its substitution into the host cation neither enlarges nor shrinks the fluorite lattice. To find r_{cr} for a given dopant cation valency, one must simply nullify the slope's corresponding term in each equation. Should a solute with an ionic radius smaller than the critical value be introduced into a fluorite oxide, the lattice will contract, and conversely, it will expand if the radius is larger.

Pyrochlore Oxides

Pyrochlore oxides, christened after the mineral pyrochlore, i.e. $(Na,Ca)_2Nb_2O_6(OH,F)$, own a crystal structure akin to the fluorite one [40].

Particularly, the $A_2B_2O_7$ pyrochlores, composed of ternary metallic oxides, possess a sufficiently intricate crystal chemistry, rendering them suitable for an expansive array of applications. Within the scope of this chapter, A and B cations are strictly regarded as possessing charges of 3^+ and 4^+, respectively. However, another full series of pyrochlore compounds involving 2^+ and 5^+ cations is also acknowledged [41].

The generic pyrochlore structure $A_2B_2O_6O'$ harbors four crystallographically distinct atomic positions, with the space group marked as Fd3m [41]. A prevalent method of outlining this structure involves anchoring its origin on the B site, positioning atoms at specific coordinates, in line with Wyckoff notation: A at 16d, B at 16c, O at 48f, and O' at 8b [41]. The pyrochlore structure's sole internal positional variable is the oxygen x parameter, a feature that describes the 48f oxygen atoms.

The pyrochlore structure is made more discernible through a convenient, fluorite-like portrayal [41, 42], as visualized in the following figure. Consequently, pyrochlore can be viewed as a systematically arranged, flawed fluorite solid solution.

Figure 2a illustrates a fraction (one-eighth) of the pyrochlore unit cell, akin to a singular fluorite unit cell, while Figure 2b, on the other hand, represents a complete pyrochlore unit cell, devoid of anions, to enhance the visual comprehension of the two cationic sublattices and their alignment along the <110> directions.

Referring to CaF_2, the fluorine anions settle in the tetrahedral locations of a Ca face-centered cubic collection. In the case of pyrochlore, the A and B cations construct the face-centered cubic arrangement and are also systematically arranged in the <110> directions. This facilitates eight-coordinate A cations and six-coordinate B cations relative to oxygen. Such cation alignment leads to a situation where tetrahedral anion sites are no longer homogenous.

More precisely, three distinct tetrahedral sites emerge in pyrochlores: the 48f, neighbored by two A and two B cations; the 8a, surrounded by four B cations; and the 8b, encircled by four A cations. Within pyrochlore, the 8a positions remain unoccupied.

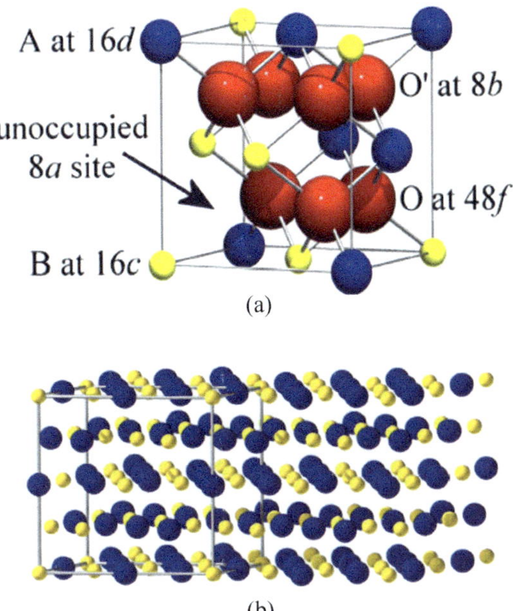

Fig. 2: (a) 1/8 of the unit cell of a generic pyrochlore $A_2B_2O_7$ (blue atoms are A^{3+} cations, yellow atoms are B^{4+} cations, red atom are O^{2-} anions); (b) complete pyrochlore sublattice.

Ionic Conduction in Fluorite Oxides

In fluorite-structured MO_2 oxides, ionic conduction takes place through the movement of oxygen ions via a vacancy mechanism [43]. This process results in the concentration of oxygen-ion vacancies increasing when solid solutions form with di- or trivalent dopant cations, thereby enhancing the overall ionic conductivity. Often, the dopant cations and oxygen-ion vacancies create defect complexes, which also play a role in conductivity [44].

The ionic conductivity in these substances, represented by u, can be described by the following equation:

$$\sigma T = A \, \exp\left(-\frac{H}{kT}\right) \tag{6}$$

Here, T stands for the absolute temperature (in K), A is a preexponential constant, and H is the activation enthalpy for conduction. Thus, at a specific temperature, conductivity escalates with a rising A and a falling H. The activation enthalpy is formulated as:

$$H = H_m + \left(\frac{n}{2}\right) H_a \tag{7}$$

In Equation (7), H_m is the enthalpy for migration of a vacancy created by doping with aliovalent cations, n equals 1 or 2 for di- and trivalent dopants correspondingly, and H_a is the enthalpy for association of a defect complex, influenced by the elastic strain field around the defect complex.

At a set temperature, the conductivity in fluorite oxides is chiefly determined by the H_m term. This term is understood to correlate with the discrepancy in ionic size between the host and dopant cations ($r_h - r_d$). Historically, the influence of diverse dopants on the ionic conductivity of fluorite-type MO_2 oxides has been explained solely through ($r_h - r_d$), meaning a dopant that reduces this difference would boost conductivity [45, 46]. However, this generalization falls short in elucidating the effects of di- and trivalent dopants on CeO_2's ionic conductivity. To this end, Kilner [46] recommended employing lattice parameter measurements to forecast dopant impacts on ionic conductivity.

From previous insights into the role of r_{cr} in defining lattice parameters, it's clear that a change in the fluorite-structure lattice due to dopant cation substitution in binary systems isn't governed by ($r_h - r_d$), but by the difference between r_{cr} and the dopant ionic radius, i.e. ($r_{cr} - r_d$). This distinction likely steers the relaxation of the lattice near the dopant-oxygen-ion vacancy defect complex.

Table 2 delineates the connection between the bulk ionic conductivity of CeO_2 doped with trivalent cations and the absolute value of ($r_{cr} - r_d$), as identified by Gerhardt and Nowick [36]. Here, the ionic radius of Ce^{4+} in eightfold coordination is given as 0.097 nm, and r_{cr} for a trivalent dopant in CeO_2 is calculated to be 0.1038 nm, as per Equation (3).

Table 2: Comparison of the relationship between the critical dopant (r_{cr}), dopant (r_d), and host cation (r_h) radii and the ionic conductivity for different cations in CeO_2 solid solution [38]

Dopant	$r_{cr} - r_d$ (nm)	$r_h - r_d$ (nm)	σ ($\times 10^{-4}$ $(\Omega\ cm)^{-1}$)
Gd3+	0.0015	0.0083	3.2
Y3+	0.0019	0.0049	2.6
La3+	0.0122	0.0190	1.9

Unlike what might be expected, Table 1 reveals that the conductivity of CeO_2 solid solutions doesn't hinge on ($r_h - r_d$), but rather on ($r_{cr} - r_d$). This observation holds true for the system CeO_2-alkaline-earth oxide as well, where r_{cr} is found to be 0.1106 nm.

Thus, the expected sequence of ionic conductivity for CeO_2 doped with alkaline-earth oxides is: CaO(0.112) > SrO(0.126) > MgO(0.089) > BaO(0.142), where the numbers in parentheses represent the ionic radius (in nm) of each cation in eightfold coordination. This pattern, within the solubility limits of the fluorite-structure phase, aligns with experimental measurements made by Yahiro et al. [37]. Interestingly, among the alkaline-earth oxides considered, MgO is responsible for the smallest ($r_h - r_d$) value.

Conversely, in the system $ZrO_2 - M_2O_3$, the anticipated order of ionic conductivity for the considered dopants is: $Yb_2O_3(0.0985) > Y_2O_3(0.1019) > Sc_2O_3(0.0870) > Sm_2O_3(0.1079)$. Again, the ionic radii (in nanometers) of the dopant cations in eightfold coordination are indicated in parentheses. For such systems, r_{cr} is determined to be 0.0948 nm using Equation (2). The obtained prediction aligns with experimental values obtained by Strickler and Carlson too, save for the case of Sc_2O_3 doping.

The empirical equations delineated earlier are applicable to systems containing multiple dopants too. Thus, the idea that maximizing conductivity is attainable through alloying with dopants that minimize the fluctuation of the fluorite lattice, has been deeply explored in literature.

In particular, Keller et al. [47] have posited an intriguing aspect of solubility regarding the lanthanide oxides in ThO_2, which depends on the absolute value of the difference in the radii of the Th^{4+} and Ln^{n+} ions. As the ionic size becomes larger, the extent of solubility increases, peaking in proximity to the radius of Nd^{3+} (0.1109 nm), after which it declines for even larger elements. Interestingly, this maximum solubility occurs with an ionic radius larger than that of Th^{4+} (0.105 nm), indicating that the behavior of solubilities in the ThO_2-Ln_2O_3 system cannot be explained solely by considering the solute cation radius.

A similar trend is observed in fluorite-structured ZrO_2 systems too [48], where the measured enthalpy has a minimum value at the Ca^{2+} radius, whilst for calculations based solely on ionic sizes MgO is expected to own the minimum value. The same contradiction is found in the CeO_2-alkaline-earth oxide system [45].

In a substitutional solid solution, the extent of solubility can be expressed as a function of both the radius difference between the solute and host element and the valency of the solute. The introduction of a cation with different radius and valency into the fluorite lattice causes strain unless the influences of the two factors are in equilibrium. This elastic energy governs the extent of solid solubility; the less energy required to introduce a dopant cation, the wider the extent of solubility for the dopant.

This deeper insight underscores the interplay between the ionic radii, valency, and elastic strain, which together determine the solubility behavior in these co-doped complex systems. In other words, it is suggested that simplistic considerations of ionic sizes alone may be insufficient to accurately predict or understand solubility trends, and emphasizes the need for a multifaceted approach that takes into account various physicochemical properties and interactions in fluorite oxides.

Even more so, the complexity of high-entropy fluorite oxides, generally proposed in literature as solid mixtures assembled in equal molar proportions from foundational materials like HfO_2, ZrO_2, and CeO_2 complemented by oxides of Y, Yb, Ca, Ti, La, Mg, and Gd acting as dopants/stabilizing agents, induces the need of developing novel approaches in estimating their final properties.

High-entropy Fluorites

In 2018, a significant discovery was made by Chen et al. [49], reporting the first entropy-stabilized fluorite oxide with the following composition: $Zr_{0.2}Hf_{0.2}Ce_{0.2}Ti_{0.2}Sn_{0.2}O_2$ with a room temperature thermal conductivity of about 1.3 $W \cdot m^{-1} K^{-1}$; and by Gild et al. [50], proposing the first series of entropy-stabilized fluorite oxide with different combinations of the five following cations: Hf, Zr, Ce, Y, Yb.

Subsequent research by various scientists led to the synthesis of different high-entropy fluorite oxides, even though almost all of these studies did not conclusively establish entropy stabilization within these systems (i.e. the reversibility from a high-temperature single-phase to a low-temperature multiphasic system). Just as an example, Djenadic et al. [51] studied equiatomic rare-earth oxides like $(Ce_{0.2}La_{0.2}Sm_{0.2}Pr_{0.2}Y_{0.2})O_2$-$\delta$, strongly contributing to the initial knowledge on high-entropy fluorite oxides.

Generally speaking, the innovative entropy-stabilized fluorite oxides ensure positive changes in enthalpy and high configurational entropy, based on variations in ionic radii, electronegativity, and the crystal structures of individual simple oxides. However, a non-intuitive structural behavior in high-entropy fluorite oxides requires a deepening in the understanding of single-phase formation, factoring in various physical, chemical and geometric parameters.

In other words, predicting unknown High Entropy Fluorite Oxides (HEFOs) that stabilize in single-phase crystal structures requires a model based on certain descriptors or indicators. Currently, efforts to model or design the stability of single-phase HEOs are in their infancy. Most research focuses on a combinatory approach of element selection and exploration of behavior in complex solid solutions [50, 52, 53]. Only a handful of papers attempt to deliberately design new high entropy oxides (not only fluorite oxides) using a clear descriptor approach, often related to geometric parameters like ionic radii. For instance: Jiang et al. [54] suggested that Goldschmidt's tolerance factor plays a role in both the formation and temperature-stability of single-phase perovskite-structured HEOs but is not the sole determinant; Wright et al. [55] proposed a descriptor for crafting high-entropy pyrochlore oxides and predicting thermal properties based on the ionic radii of individual elements and their average value.

Specifically for HEFOs, Spiridigliozzi et al. recently correlated cationic radii and statistical properties connected to them to develop a predictive model [56, 57]. As HEFOs include different lanthanides, bixbyite-like structures (such as Gd_2O_3 or Ho_2O_3) have been considered as possibly competing with the fluorite structure formation. Given that, it has been found that reducing the size mismatch among cations of a generic HEFO would favor bixbyite formation, and vice versa. In other words, as the most straightforward way to measure this mismatch is the standard deviation (σ) of the cationic radii distribution, a clear and definite trend emerged:

- when $\sigma > 0.095$, a fluorite-like single phase HEO is obtained;

- when $\sigma \approx 0.095$, a two-phase (fluorite + bixbyite) system is obtained;
- when $\sigma < 0.095$, a bixbyite-like single-phase HEO is obtained.*

Thus, a fluorite-like structure is encouraged when there is greater "disorder" in a given system, stemming from a more significant mismatch in the cation sizes, leading to higher internal lattice stresses. Conversely, a bixbyite-like structure is preferred when these stresses are lower, likely because it is better able to stably accommodate such stresses.

Other than obtaining or not a single-phase HEFO, it is crucial to deepen the understanding of the order-disorder transition (ODT) of these extremely complex fluorite oxides, as gaining a deeper understanding and control over these ordering and disordering mechanisms across different scales might pave the way to fine-tune a plethora of interesting physical properties of fluorite oxides.

Many studies have delved into this aspect for conventional fluorite oxides, by using techniques such as X-ray diffraction [58, 59], Raman spectroscopy [60], transmission electron microscopy [61, 62], X-ray absorption near edge spectroscopy (XANES) [63], and neutron total scattering [64, 65]. Notably, researchers have pinpointed the primary reason for the long-range transition from fluorite to pyrochlore in 2- or 3-component oxides as the oxygen anion's shift from 48f to 8a positions [58, 66]. Furthermore, there have been reports of weberite-type short-range ordering within the long-range-disordered defect fluorite structure [67, 68]. For instance, a study by Drey et al. [69] postulated that the movement of oxygen anions can lead to the development of seven-coordinated Zr sites, resulting in weberite-type short-range order in $Ho_2(Ti_{2-y}Zr_y)O_7$.

Very recently, Zhang et al. [70] delved into the intricate relationship between long and short-range ordering and disordering in a 10-component high-entropy system, by using neutron diffraction and neutron total scattering techniques. The compound that they synthesized displayed high-entropy phases that had either a defect fluorite or pyrochlore long-range order. Interestingly, the long-range order-to-disorder transition in Zhang's HEOs didn't reveal any dual or multiple phase areas, contrasting with simpler oxide structures. The stability criteria for the pyrochlore vs fluorite phase in such 10-component systems differ from those established for simpler (with 5 different cations) HEO systems. In particular, it has been found that certain local polyhedral environments remained consistent in specific samples near the order-to-disorder transition.

In terms of practical implications, Zhang et al. demonstrated that room temperature thermal conductivity of 10-component high-entropy systems showed minimal change during the long-range transition, but a significant reduction was observed with the emergence of short-range weberite-type ordering. This suggests that short-range structural changes can more effectively reduce thermal conductivity than the broader long-range transitions.

Ultimately, understanding and manipulating these short and long-range orders and disorders in complex HEFOs could pave the way to customize their structure and properties for different applications, thermal barrier coatings and catalysts being the most promising ones.

High-Entropy Fluorites as Thermal Barrier Coatings

Thermal barrier coatings (TBCs) are crucial components in the hot end parts of gas turbines. These parts have now reached gas inlet temperatures that surpass the maximum usage temperature of nickel-based superalloys, which is no more than 1150 °C, and these alloys are employed in manufacturing turbine blades and other hot end elements [71]. To shield these superalloys from extreme operating temperatures, TBCs are being extensively explored and utilized.

6–8 wt% yttria-stabilized zirconia (YSZ) has become a commercial choice for TBCs, thanks to its relatively low thermal conductivity of 2.40 $W \cdot m^{-1} \cdot K^{-1}$ at 1000 °C [72], high thermal expansion coefficient (TEC) of 10.9×10^{-6} K^{-1} [73], impressive fracture toughness of about 3 $MPa \cdot$ [74], and low Young's modulus of approximately 240 GPa [75]. However, the use of YSZ is severely limited due to a phase transformation that occurs during prolonged service above 1200 °C. This transformation leads to significant volume changes in the material, causing the TBCs to fail and spall [76-78]. Consequently, the urgent need to discover new types of TBCs with improved temperature phase stability and reduced thermal conductivity has arisen.

Three primary strategies have been developed over recent decades to address these issues:

- the addition of rare-earth doping elements to YSZ, such as gadolinia (Gd_2O_3) [79], neodymia (Nd_2O_3) [80], erbia (Er_2O_3) [81], and ytterbia (Yb_2O_3) [80], to hinder its phase transformation at high temperatures;
- the exploration of alternative advanced materials for TBCs, like $La_2Ce_2O_7$ [82], $Gd_2Zr_2O_7$ [83], and $DyTaO_4$ [84];
- the fabrication of multi-layered TBC top coats using cutting-edge thermal spray techniques, including atmospheric plasma spraying, suspension plasma spraying, and solution plasma spraying [85, 86].

However, despite extensive research aimed to enhance the performance of conventional YSZ, fulfilling the growing requirements of modern high-efficiency gas turbines and jets remains an open challenge.

In the search for novel high-performing systems for TBCs, high-entropy fluorite oxides have proven to be promising candidates, displaying lower thermal conductivity compared to corresponding constituent materials [87-89]. Their superior thermal insulation characteristic largely stems from the "cocktail effect", i.e. an attribute of compositionally complex compounds, where the combined impact of multi-component elements leads to unexpected enhancements in their thermo-physical properties. Additionally, high-entropy oxides provide exceptional phase stability and a slow grain growth rate due to the high entropy effect and sluggish diffusion [90, 91].

For TBCs, excellent thermal insulation is paramount. The thermal conductivity of high entropy oxides is primarily governed by phonon scattering, which encompasses intrinsic scattering between phonons, lattice defect scattering, grain boundary scattering in polycrystalline materials, and so forth. Moreover, the unique arrangement of multi-component cations in high-entropy oxides creates considerable variances in atomic mass, ionic radius, and interatomic force, resulting in serious lattice distortion and a considerable increase in phonon scattering. As a consequence, their overall thermal conductivity is substantially reduced compared to their "parent" structures.

Among the most notable literature findings about high-entropy oxides with very low thermal conductivities are the following: Chen et al. [49] unveiled an entropy-stabilized fluorite oxide with room temperature thermal conductivity as low as 1.28 $W \cdot m^{-1} \cdot K^{-1}$; Zhao et al. discovered a novel high-entropy ceramic exhibiting a thermal conductivity of only 0.76 $W \cdot m^{-1} \cdot K^{-1}$ at room temperature and a slow grain growth rate, making it apt for thermal barrier coatings [92].

Moreover, the high entropy $(La_{0.2}Nd_{0.2}Yb_{0.2}Y_{0.2}Lu_{0.2})_2Ce_2O_7$ system exhibited low thermal diffusivity, ranging from 0.65 to 0.31 mm²/s, from room temperature up to 1000 °C [93], whilst another system, $(La_{1/6}Nd_{1/6}Yb_{1/6}Y_{1/6}Sm_{1/6}Lu_{1/6})_2Ce_2O_7$, showed lower thermal conductivity than Sm_2Ce_2O [94]. The $(Hf_{0.25}Zr_{0.25}Ce_{0.25}Y_{0.25})O_{1.875}$ entropy-stabilized fluorite oxide owns lower thermal conductivities and hardness compared to 8YSZ [50].

Thus, these multi-component rare-earth high-entropy fluorite/pyrochlore oxides reveal promising potential for applications in the thermal expansion materials sector, but suitable structural and performance optimizations still need to be performed.

The introduction of specific rare earth elements in such systems can enhance the mechanical and thermomechanical properties of high-entropy zirconates, cerates, and tantalates by altering phonon scattering, complex element compositions, and lattice distortions [95, 96].

In the recent work of Guo et al. [97], two types of high-entropy fluorite oxides were synthesized: $(Hf_{0.25}Zr_{0.25}Ce_{0.25}Y_{0.25})O_{1.875}$ and $(Zr_{0.2}Ce_{0.2}Hf_{0.2}Y_{0.2}RE_{0.2})O_{1.8}$ (where RE = La, Nd, Sm). The study investigated the phase structure, mechanical and thermophysical properties of these ceramics, and explored the impact of each rare earth element on the structure and properties of such high-entropy systems.

Generally speaking, the low thermal conductivity in HEFOs can be traced back to two main factors. Firstly, HEFOs contain a large volume of oxygen vacancies that act as phonon scattering sites. Secondly, the chemical complexity within HEFOs results in fluctuations of mass and strain fields, reducing phonon propagation and leading to additional scattering, thus impacting the thermal conductivities of HEFOs [98, 99].

Other than thermal insulation, the thermal expansion coefficient (TEC) is vitally important for TBC too. YSZ, with its relatively low TECs, may cause a significant thermal mismatch between TBC materials and metallic substrates, potentially leading to coating failure [100]. Dilatometric measurements of a recent study [101] indicate that the average TECs of HEFOs within the temperature span of 673 K to 1100 K fall within the range of 11.9×10^{-6}-12.1×10^{-6} K^{-1}, closely aligning with $La_2Ce_2O_7$ (12.3×10^{-6} K^{-1}) and exceeding YSZ (10.9×10^{-6} K^{-1}).

This high thermal expansion coefficient can be attributed to the substantial size of Ce^{4+} and the fragile Ce-O bonds [102]. This factor not only assists in minimizing thermal expansion discrepancies between TBCs and metallic substrates but also ensures a strong linkage between coating materials and metallic substrates upon usage. According to solid thermal expansion theory [103], the TEC is directly related to the ionic bond strength of the crystal components, determined by the mean distance between atoms within the lattice structure.

Moreover, within the temperature range of 473 K to 673 K, the HEFOs' TECs can be adjusted from the negative value of $La_2Ce_2O_7$ to positive ones, resulting in superior TEC alignment with metal substrates. The HEFOs also exhibit remarkable resistance to sintering and maintain phase stability at extremely high temperatures, even up to 1873 K.

The blend of compatible TECs with metal substrates, reduced thermal conductivities, strong sintering resistance, and stability at elevated temperatures positions HEFOs as the most promising option for next-generation thermal barrier coatings and thermally insulating materials.

High-Entropy Fluorites as Catalysts

Recent studies have highlighted the potential of high-entropy oxides in diverse energy-related applications, with catalysis standing out prominently [104-106]. Distinct from doped transition metal oxides and low or medium-entropy oxides, HEOs boast a disordered chemical composition and atomic distribution, their combination granting them superior qualities in terms of thermodynamics, crystal structure, and kinetics.

Therefore, the exceptional catalytic performance of HEOs is often shaped by their unique architectural traits. The confinement of cations with different radii within the same lattice, thanks to increased configurational entropy, leads to substantial internal lattice distortion. This major distortion creates oxygen defects, raises the quantity of oxygen vacancies at the active catalytic center, and reduces the system's energy. This combination of effects helps in the activation and movement of active species [107]. Moreover, the kinetic hysteresis diffusion effect within HEOs enhances their thermal stability in catalytic processes [108,109].

High-entropy ceria-based fluorite oxides have recently caught the interest of researchers for their interesting catalytic activity, as Zhang et al. [110] introduced a mechanochemical technique to develop heterostructured, highly stable CuCeOx-HEO catalysts. Notably, CuCeOx-HEO catalysts showcased impressive resistance to high temperatures, with almost no noticeable activity loss post-calcination at 900 °C.

In a parallel approach, Shu et al. [111] managed to surpass the natural solubility constraints of aliovalent metal ions in the fluorite configuration, as they successfully integrated 10 mol% each of Cu^{2+}, Co^{2+}, Mg^{2+}, Ni^{2+}, and Zn^{2+} ions into the CeO_2 fluorite-like matrix. This enhanced the generation of vacancy-driven reactive oxygen entities and induced local lattice modifications. Thanks to a plethora of surface sites that minimize the competitive absorption of varied gas components, these high-entropy ceria-based catalysts enabled effective processing of CO-toluene-propylene and standard car exhaust, exhibiting superior heat and hydrothermal resistance.

Finally, Xu et al. [112] showcased the entropy-fueled stability of Pd single-atom catalysts (SACs) using high-entropy fluorite oxides as support. In their study, the researchers employed a combination of mechanical milling and intense heat treatment to synthesize several $(Pd_yCeZrHfTiLa)O_x$ solid mixtures in which Pd established Pd—O—M connections (where M stands for Ce, Zr, Hf, Ti, and La) by replacing cations in the fluorite lattice. Catalytic testing on such system revealed heightened surface adaptability and an increase in oxygen vacancies, aiding the adsorption and triggering of molecular oxygen, leading to improved CO oxidation processes. Notably, the novel Pd@HEFO catalyst showcased remarkable heat and moisture resistance, especially in the concurrent oxidation of CO, C_3H_6, and NO, underscoring its feasibility for purifying diesel engine emissions.

To conclude, it is important to note that the field of designing HEOs-based materials for catalytic applications is still in its infancy, even though the opportunity to fine-tune their structures to cater to specific catalytic requirements make them promising for future applications.

References

[1] Catlow, C.R.A. (1984). Transport in doped fluorite oxides. Solid State Ionics, 12, 67-73.

[2] Krivovichev, S.V. (1999). Systematics of fluorite-related structures. I. General Principles. Solid State Sciences, 1(4), 211-219.

[3] Marlton, F.P., Zhang, Z., Zhang, Y., Proffen, T.E., Ling, C.D. & Kennedy, B.J. (2021). Lattice disorder and oxygen migration pathways in pyrochlore and defect-fluorite oxides. Chemistry of Materials, 33(4), 1407-1415.

[4] Sattonnay, G., Moll, S., Desbrosses, V., Bekale, V.M., Legros, C., Thomé, L. & Monnet, I. (2010). Mechanical properties of fluorite-related oxides subjected to swift ion irradiation: Pyrochlore and zirconia. Nuclear Instruments and Methods in Physics Research Section B: Beam Interactions with Materials and Atoms, 268(19), 3040-3043.

[5] Wang, C., Wang, Y., Zhang, A., Cheng, Y., Chi, F. & Yu, Z. (2013). The influence of ionic radii on the grain growth and sintering-resistance of $Ln_2Ce_2O_7$ (Ln = La, Nd, Sm, Gd). Journal of Materials Science, 48, 8133-8139.

[6] Navrotsky, A. (2010). Thermodynamics of solid electrolytes and related oxide ceramics based on the fluorite structure. Journal of Materials Chemistry, 20(47), 10577-10587.

[7] Tuller, H.L. (1992). Mixed ionic-electronic conduction in a number of fluorite and pyrochlore compounds. Solid State Ionics, 52(1-3), 135-146.

[8] Xu, Q., Pan, W., Wang, J., Wan, C., Qi, L., Miao, H. & Torigoe, T. (2006). Rare-earth zirconate ceramics with fluorite structure for thermal barrier coatings. Journal of the American Ceramic Society, 89(1), 340-342.

[9] Deijkers, J.A., Begley, M.R. & Wadley, H.N. (2022). Failure mechanisms in model thermal and environmental barrier coating systems. Journal of the European Ceramic Society, 42(12), 5129-5144.

[10] Sadovskaya, E.M., Ivanova, Y.A., Pinaeva, L.G., Grasso, G., Kuznetsova, T.G., van Veen, A. & Mirodatos, C. (2007). Kinetics of oxygen exchange over CeO_2-ZrO_2 fluorite-based catalysts. The Journal of Physical Chemistry A, 111(20), 4498-4505.

[11] Birkby, I. & Stevens, R. (1996). Applications of zirconia ceramics. Key Engineering Materials, 122, 527-552.

[12] Sameshima, S., Ichikawa, T., Kawaminami, M. & Hirata, Y. (1999). Thermal and mechanical properties of rare earth-doped ceria ceramics. Materials Chemistry and Physics, 61(1), 31-35.

[13] Kumar, V.P., Reddy, Y.S., Kistaiah, P., Prasad, G. & Reddy, C.V. (2008). Thermal and electrical properties of rare-earth co-doped ceria ceramics. Materials Chemistry and Physics, 112(2), 711-718.

[14] Li, C., Ren, C., Ma, Y., He, J. & Guo, H. (2021). Effects of rare earth oxides on microstructures and thermo-physical properties of hafnia ceramics. Journal of Materials Science & Technology, 72, 144-153.

[15] Alotaibi, M., Li, L. & West, A.R. (2021). Electrical properties of yttria-stabilised hafnia ceramics. Physical Chemistry Chemical Physics, 23(45), 25951-25960.

[16] Pauling, L. The Nature of the Chemical Bond and the Structure of Molecules and Crystals: An Introduction to Modern Structural Chemistry. Cornell University Press, Ithaca, NY.

[17] Stanek, (August 2003). Atomic Scale Disorder in Fluorite and Fluorite Related Oxides. Department of Materials, Imperial College of Science, Technology and Medicine.

[18] Chiang, Y.M., Birnie III, D.P. & Kingery, W.D. Physical Ceramics. John Wiley & Sons, Inc., New York.

[19] Shannon, R.D. (1976). Revised effective ionic radii and systematic studies of interatomic distances in halides and chalcogenides. Acta Crystallographica A, 32(5), 751-767.

[20] Mogensen, M., Lybye, D., Bonanos, N., Hendriksen, P.V. & Poulsen, F.W. (2004). Factors controlling the oxide ion conductivity of fluorite and perovskite structured oxides. Solid State Ionics, 174(1-4), 279-286.

[21] Wachsman, E.D. (2004). Effect of oxygen sublattice order on conductivity in highly defective fluorite oxides. Journal of the European Ceramic Society, 24(6), 1281-1285.

[22] Clarke, D.R. & Phillpot, S.R. (2005). Thermal barrier coating materials. Materials Today, 8(6), 22-29.

[23] Song, D., Kim, J., Lyu, G., Pyeon, J., Jeon, H.B., Oh, Y.S. & Jung, Y.G. (2021). Effect of cation substitution on thermophysical properties of fluorite A_3BO_7 ceramics. Journal of Alloys and Compounds, 883, 160848.

[24] Wirkus, C.D. & Berard, M.F. (1982). Abradability and hardness in rare-earth-oxide stabilized hafnia. Journal of Materials Science, 17, 109-114.

[25] Navrotsky, A. (2010). Thermodynamics of solid electrolytes and related oxide ceramics based on the fluorite structure. Journal of Materials Chemistry, 20(47), 10577-10587.

[26] Izu, N., Nishizaki, S., Shin, W., Itoh, T., Nishibori, M. & Matsubara, I. (2009). Resistive oxygen sensor using ceria-zirconia sensor material and ceria-yttria temperature compensating material for lean-burn engine. Sensors, 9(11), 8884-8895.

[27] Dong, Y., Sun, X., Mu, A., Liu, Z., Qiu, G., Zhang, X., ... & Wang, X. (2022). Sensing property of limiting current oxygen sensor with (8YSZ) 0.4 (CeO$_2$) 0.6 dense diffusion barrier based on 8YSZ solid electrolyte. Materials Science in Semiconductor Processing, 143, 106528.

[28] Nikonov, A., Pavzderin, N. & Khrustov, V. (2022). Dense electrode layers-supported microtubular oxygen pump. Membranes, 12(11), 1114.

[29] Liu, T., Zhang, X., Wang, X., Yu, J. & Li, L. (2016). A review of zirconia-based solid electrolytes. Ionics, 22, 2249-2262.

[30] Cao, J., Su, C., Ji, Y., Yang, G. & Shao, Z. (2021). Recent advances and perspectives of fluorite and perovskite-based dual-ion conducting solid oxide fuel cells. Journal of Energy Chemistry, 57, 406-427.

[31] Cao, J., Su, C., Ji, Y., Yang, G. & Shao, Z. (2021). Recent advances and perspectives of fluorite and perovskite-based dual-ion conducting solid oxide fuel cells. Journal of Energy Chemistry, 57, 406-427.

[32] Zhang, S., Savaniu, C. & Irvine, J.T. (2019). Fluorite materials for SOFC electrolyte applications. ECS Transactions, 91(1), 1111.

[33] Spiridigliozzi, L., Dell'Agli, G., Accardo, G., Yoon, S.P. & Frattini, D. (2019). Electro-morphological, structural, thermal and ionic conduction properties of Gd/Pr co-doped ceria electrolytes exhibiting mixed Pr^{3+}/Pr^{4+} cations. Ceramics International, 45(4), 4570-4580.

[34] Glushkova, V.B., Hanic, F. & Sazonova, L.V. (1978). Lattice parameters of cubic solid solutions in the systems uR2O$_3$-(1-u)MO$_2$. Ceramurgia International, 4(4), 176-178.

[35] Ingel, R.P. & III, D.L. (1986). Lattice parameters and density for Y_2O_3-stabilized ZrO_2. Journal of the American Ceramic Society, 69(4), 325-332.

[36] Gerhardt, R. & Nowick, A.S. (1986). Grain-boundary effect in ceria doped with trivalent cations: I, Electrical measurements. Journal of the American Ceramic Society, 69(9), 641-646.

[37] Yahiro, H., Ohuchi, T., Eguchi, K. & Arai, H. (1988). Electrical properties and microstructure in the system ceria-alkaline earth oxide. Journal of Materials Science, 23, 1036-1041.

[38] Kim, D.J. (1989). Lattice parameters, ionic conductivities, and solubility limits in fluorite-structure MO$_2$ oxide [M = Hf^{4+}, Zr^{4+}, Ce^{4+}, Th^{4+}, U^{4+}] solid solutions. Journal of the American Ceramic Society, 72(8), 1415-1421.

[39] Denton, A.R. & Ashcroft, N.W. (1991). Vegard's law. Physical Review A, 43(6), 3161.

[40] Atencio, D., Andrade, M.B., Christy, A.G., Gieré, R. & Kartashov, P.M. (2010). The pyrochlore supergroup of minerals: Nomenclature. The Canadian Mineralogist, 48(3), 673-698.

[41] Subramanian, M.A., Aravamudan, G. & Rao, G.S. (1983). Oxide pyrochlores—Areview. Progress in Solid State Chemistry, 15(2), 55-143.

[42] Aleshin, E. & Roy, R. (1962). Crystal chemistry of pyrochlore. Journal of the American Ceramic Society, 45(1), 18-25.

[43] Subbarao, E.C. & Maiti, H.S. (1984). Solid electrolytes with oxygen ion conduction. Solid State Ionics, 11(4), 317-338.

[44] Hohnke, D.K. (1981). Ionic conduction in doped oxides with the fluorite structure. Solid State Ionics, 5, 531-534.

[45] Catlow, C.R.A. (1984). Transport in doped fluorite oxides. Solid State Ionics, 12, 67-73.

[46] Kilner, J.A. (1982). The role of dopant size in determining oxygen ion conductivity in the fluorite structure oxides. Studies in Inorganic Chemistry, 3, 189-192.

[47] Keller, C., Berndt, U., Engerer, H. & Leitner, L. (1972). Phasengleichgewichte in den systemen thoriumoxid-lanthanidenoxide. Journal of Solid State Chemistry, 4(3), 453-465.

[48] Mackrodt, W.C. & Woodrow, P.M. (1986). Theoretical estimates of point defect energies in cubic zirconia. Journal of the American Ceramic Society (United States), 69(3).

[49] Chen, K., Pei, X., Tang, L., Cheng, H., Li, Z., Li, C., ... & An, L. (2018). A five-component entropy-stabilized fluorite oxide. Journal of the European Ceramic Society, 38(11), 4161-4164.

[50] Gild, J., Samiee, M., Braun, J.L., Harrington, T., Vega, H., Hopkins, P.E. & Luo, J. (2018). High-entropy fluorite oxides. Journal of the European Ceramic Society, 38(10), 3578-3584.

[51] Djenadic, R., Sarkar, A., Clemens, O., Loho, C., Botros, M., Chakravadhanula, V.S., ... & Hahn, H. (2017). Multicomponent equiatomic rare earth oxides. Materials Research Letters, 5(2), 102-109.

[52] Dąbrowa, J., Szymczak, M., Zajusz, M., Mikuła, A., Moździerz, M., Berent, K. & Świerczek, K. (2020). Stabilizing fluorite structure in ceria-based high-entropy oxides: Influence of Mo addition on crystal structure and transport properties. Journal of the European Ceramic Society, 40(15), 5870-5881.

[53] Zhang, R.Z. & Reece, M.J. (2019).Review of high entropy ceramics: Design, synthesis, structure and properties, Journal of Materials Chemistry A, 7, 22148 .

[54] Jiang, S., Hu, T., Gild, J., Zhou, N., Nie, J., Qin, M., ... & Luo, J. (2018). A new class of high-entropy perovskite oxides. Scripta Materialia, 142, 116-120 .

[55] Wright, A.J., Wang, Q., Ko, S.T., Chung, K.M., Chen, R. & Luo, J. (2020). Size disorder as a descriptor for predicting reduced thermal conductivity in medium- and high-entropy pyrochlore oxides. Scripta Materialia, 181, 76-81.

[56] Spiridigliozzi, L., Ferone, C., Cioffi, R. & Dell'Agli, G. (2021). A simple and effective predictor to design novel fluorite-structured high entropy oxides (HEOs). Acta Materialia, 202, 181-189.

[57] Spiridigliozzi, L., Ferone, C., Cioffi, R. & Dell'Agli, G. (2023). Compositional design of single-phase rare-earth based high-entropy oxides (HEOs) by using the cluster-plus-glue atom model. Ceramics International, 49(5), 7662-7669.

[58] Zhang, Z., Middleburgh, S.C., De Los Reyes, M., Lumpkin, G.R., Kennedy, B.J., Blanchard, P.E.R., ... & Reynolds, E. (2013).Gradual structural evolution from pyrochlore to defect-fluorite in $Y_2Sn_{2-x}Zr_xO_7$: Average vs local structure. Journal of Physical Chemistry C, 117, 26740-26749.

[59] Maram, P.S., Ushakov, S.V., Weber, R.J.K., Benmore, C.J. & Navrotsky, A. (2018). Probing disorder in pyrochlore oxides using in situ synchrotron diffraction from levitated solids – A thermodynamic perspective. Scientific Reports, 8, 1-11.

[60] Zhou, L., Huang, Z., Qi, J., Feng, Z., Wu, D., Zhang, W., ... & Lu, T. (2016). Thermal-driven fluorite-pyrochlore-fluorite phase transitions of $Gd_2Zr_2O_7$ ceramics probed in large range of sintering temperature. Metallurgical and Materials Transactions A, 47, 623-630.

[61] O'Quinn, E.C., Tracy, C.L., Cureton, W.F., Sachan, R., Neuefeind, J.C., Trautmann, C. & Lang, M.K. (2021). Multi-scale investigation of heterogeneous swift heavy ion tracks in stannate pyrochlore. Journal of Materials Chemistry A, 9, 16982-16997.

[62] Karthik, C., Anderson, T.J., Gout, D. & Ubic, R. (2012). Transmission electron microscopic study of pyrochlore to defect-fluorite transition in rare-earth pyrohafnates. Journal of Solid State Chemistry, 194, 168-172.

[63] Blanchard, P.E.R., Clements, R., Kennedy, B.J., Ling, C.D., Reynolds, E., Avdeev, M., ... & Jang, L.Y. (2012). Does local disorder occur in the pyrochlore zirconates? Inorganic Chemistry, 51, 13237-13244.

[64] Shafique, M., Kennedy, B.J., Iqbal, Y. & Ubic, R. (2016). The effect of B-site substitution on structural transformation and ionic conductivity in Ho2(ZryTi1-y)$_2$O$_2$. J. Alloys Compd., 671, 226-233.

[65] Norberg, S.T., Hull, S., Eriksson, S.G., Ahmed, I., Kinyanjui, F. & Biendicho, J.J. (2012). Pyrochlore to fluorite transition: The Y_2(Ti1-xZrx)$_2$O$_7$ ($0.0 \leq x \leq 1.0$) system. Chemistry of Materials, 24, 4294-4300.

[66] Clements, R., Hester, J.R., Kennedy, B.J., Ling, C.D. & Stampfl, A.P.J. (2011). The fluorite pyrochlore transformation of Ho$_2$-yNdyZr$_2$O$_2$. Journal of Solid State Chemistry, 184, 2108-2113.

[67] Shamblin, J., Tracy, C.L., Palomares, R.I., O'Quinn, E.C., Ewing, R.C., Neuefeind, J., ... & Lang, M. (2018). Similar local order in disordered fluorite and aperiodic pyrochlore structures. Scripta Materialia, 144, 60-67.

[68] Sherrod, R., O'Quinn, E.C., Gussev, I.M., Overstreet, C., Neuefeind, J. & Lang, M.K. (2021). Comparison of short-range order in irradiated dysprosium titanates. npj Materials Degradation, 5, 1-7.

[69] Drey, D.L., O'Quinn, E.C., Subramani, T., Lilova, K., Baldinozzi, G., Gussev, I.M., ... & Lang, M. (2020). Disorder in Ho$_2$Ti$_2$-xZrxO$_7$: Pyrochlore to defect fluorite solid solution series. RSC Advances, 10, 34632-34650.

[70] Zhang, D., Chen, Y., Vega, H., Feng, T., Yu, D., Everett, M., ... & Luo, J. (2023). Long- and short-range orders in 10-component compositionally complex ceramics. Advanced Powder Materials, 2(2), 100098.

[71] Darolia, R.(2013). Thermal barrier coatings technology: Critical review, progress update, remaining challenges and prospects. International Materials Reviews, 58, 315e48.

[72] Zhao, M., Ren, X., Yang, J. & Pan, W. (2016). Thermo-mechanical properties of ThO2-doped Y2O3 stabilized ZrO2 for thermal barrier coatings. Ceramics International, 42(1), 501-508.

[73] Cao, X.Q., Vassen, R., Tietz, F. & Stoever, D. (2006). New double-ceramic-layer thermal barrier coatings based on zirconia–rare earth composite oxides. Journal of the European ceramic society, 26(3), 247-251.

[74] Dwivedi, G., Viswanathan, V., Sampath, S., Shyam, A. & Lara-Curzio, E. (2014). Fracture toughness of plasma-sprayed thermal barrier ceramics: influence of processing, microstructure, and thermal aging. Journal of the American Ceramic Society, 97(9), 2736-2744.

[75] Ren, X.R. and Pan, W. (2014). Mechanical properties of high-temperature degraded yttria-stabilized zirconia. Acta Materialia, 69, 397-406.

[76] Xing, C., Yi, M.Y., Shan, X. et al.(2020). Sintering behavior of a nanostructured thermal barrier coating deposited using electro-sprayed particles. Journal of the American Ceramic Society, 103, 7267-7282.

[77] Limarga, A.M., Shian, S., Baram, M. & Clarke, D.R. (2012). Effect of high-temperature aging on the thermal conductivity of nanocrystalline tetragonal yttria-stabilized zirconia. Acta Materialia, 60(15), 5417-5424.

[78] Tan, Y., Longtin, J. P., Sampath, S. & Wang, H. (2009). Effect of the starting microstructure on the thermal properties of as-sprayed and thermally exposed plasma-sprayed YSZ coatings. Journal of the American Ceramic Society, 92(3), 710-716.

[79] Wang, Y.X. & Zhou, C.G. (2017). Hot corrosion behavior of nanostructured Gd_2O_3 doped YSZ thermal barrier coating in presence of Na_2SO_4 + V_2O_5 molten salts. Progress in Natural Science: Materials International, 27, 507-513.

[80] Wei, Q.L., Guo, H.B. and Gong, S.K. (2007). Microstructure evolution of Nd_2O_3 and Yb_2O_3 co-doped YSZ thermal barrier coatings during high temperature exposure. Materials Science Forum, 546-549, 1735-1738.

[81] Wang, Q., Guo, L., Yan, Z. & Ye, F. (2018). Phase composition, thermal conductivity, and toughness of TiO2-doped, Er2O3-stabilized ZrO2 for thermal barrier coating applications. Coatings, 8(7), 253.

[82] Cao, X., Vassen, R., Fischer, W., Tietz, F., Jungen, W. & Stöver, D. (2003). Lanthanum–cerium oxide as a thermal barrier-coating material for high-temperature applications. Advanced Materials, 15(17), 1438-1442.

[83] Shen, Z., Liu, G., Mu, R., He, L., Xu, Z. & Dai, J. (2021). Effects of Er stabilization on thermal property and failure behavior of Gd2Zr2O7 thermal barrier coatings. Corrosion Science, 185, 109418.

[84] Wang, J., Chong, X., Zhou, R. & Feng, J. (2017). Microstructure and thermal properties of RETaO4 (RE= Nd, Eu, Gd, Dy, Er, Yb, Lu) as promising thermal barrier coating materials. Scripta Materialia, 126, 24-28.

[85] Lashmi, P.G., Ananthapadmanabhan, P.V., Unnikrishnan, G. & Aruna, S.T. (2020). Present status and future prospects of plasma sprayed multilayered thermal barrier coating systems. Journal of the European Ceramic Society, 40(8), 2731-2745.

[86] Kebriyaei, A., Rahimipour, M.R., Razavi, M. and Alizade Herfati, A. (2023). Effect of solution precursor on microstructure and high-temperature properties of the thermal barrier coating made by solution precursor plasma spray (SPPS) process. Journal of Thermal Spray Technology, 32(1), 8-28.

[87] Cong, L., Li, W., Wang, J., Gu, S. and Zhang, S. (2022). High-entropy $(Y_{0.2}Gd_{0.2}Dy_{0.2}Er_{0.2}Yb_{0.2})$ $2Hf_2O_7$ ceramic: A promising thermal barrier coating material. Journal of Materials Science & Technology, 101, 199-204.

[88] Cong, L., Gu, S. and Li, W. (2021). Thermophysical properties of a novel high entropy hafnate ceramic. Journal of Materials Science & Technology, 85, 152-157.

[89] Ren, K., Wang, Q., Shao, G., Zhao, X. & Wang, Y. (2020). Multicomponent high-entropy zirconates with comprehensive properties for advanced thermal barrier coating. Scripta Materialia, 178, 382-386.

[90] Sun, J., Guo, L., Zhang, Y., Wang, Y., Fan, K. & Tang, Y. (2022). Superior phase stability of high entropy oxide ceramic in a wide temperature range. Journal of the European Ceramic Society, 42(12), 5053-5064.

[91] Ping, X., Meng, B., Li, C., Lin, W., Chen, Y., Fang, C. & Zheng, Q. (2022). Thermophysical and electrical properties of rare-earth-cerate high-entropy ceramics. Journal of the American Ceramic Society, 105(7), 4910-4920.

[92] Zhao, Z., Xiang, H., Dai, F.Z., Peng, Z. & Zhou, Y. (2019). $(La_{0.2}Ce_{0.2}Nd_{0.2}Sm_{0.2}Eu_{0.2})$ $2Zr_2O_7$: A novel high-entropy ceramic with low thermal conductivity and sluggish grain growth rate. Journal of Materials Science & Technology, 35(11), 2647-2651.

[93] Tang, A., Li, B., Sang, W., Hongsong, Z., Chen, X., Zhang, H. & Ren, B. (2022). Thermophysical performances of high-entropy $(La_{0.2}Nd_{0.2}Yb_{0.2}Y_{0.2}Sm_{0.2})2Ce_2O_7$ and $(La_{0.2}Nd_{0.2}Yb_{0.2}Y_{0.2}Lu_{0.2})2Ce_2O_7$ oxides. Ceramics International, 48(4), 5574-5580.

[94] Zhang, H., Zhao, L., Sang, W., Chen, X., Tang, A. & Zhang, H. (2022). Thermophysical performances of $(La_1/6Nd_1/6Yb_1/6Y_1/6Sm_1/6Lu_{1/6})_2Ce_2O_7$ high-entropy ceramics for thermal barrier coating applications. Ceramics International, 48, 1512-1521.

[95] Li, C., Meng, B., Fan, S., Ping, X., Fang, C., Lin, W. et al. (2022). Rare-earth tantalate high-entropy ceramics with sluggish grain growth and low thermal conductivity. Ceramics International, 48, 11124-11133.

[96] Liu, D., Wang, Y., Zhou, F., Xu, B. & Lv, B. (2021). A novel high-entropy $(Sm_{0.2}Eu_{0.2}Tb_{0.2}Dy_{0.2}Lu_{0.2})_2Zr_2O_7$ ceramic aerogel with ultralow thermal conductivity. Ceramics International, 47, 29960-29968.

[97] Guo, X., Yu, Y., Ma, W., Tang, H., Qiao, Z., Zhou, F. & Liu, W. (2022). Thermal properties of $(Zr_{0.2}Ce_{0.2}Hf_{0.2}Y_{0.2}RE_{0.2})$ $O_{1.8}$ (RE = La, Nd and Sm) high entropy ceramics for thermal barrier materials. Ceramics International, 48(24), 36084-36090.

[98] Wan, C., Qu, Z., Du, A. & Pan, W. (2009). Influence of B site substituent Ti on the structure and thermophysical properties of $A_2B_2O_7$-type pyrochlore $Gd_2Zr_2O_7$. Acta Materialia, 57(16), 4782-4789.

[99] Wan, C., Pan, W., Xu, Q., Qin, Y., Wang, J., Qu, Z. & Fang, M. (2006). Effect of point defects on the thermal transport properties of $(LaxGd1-x)_2Zr_2O_7$: Experiment and theoretical model. Physical Review B, 74(14), 144109.

[100] Jiang, T., Xie, M., Wang, X. & Song, X. (2020). Effects of Nb^{5+} doping on thermal properties of $Gd_2(Zr1-xNbx)_2O_7+x$ ceramics. Advances in Applied Ceramics, 119, 212-217.

[101] Xu, L., Wang, H., Su, L., Lu, D., Peng, K. & Gao, H. (2021). A new class of high-entropy fluorite oxides with tunable expansion coefficients, low thermal conductivity and exceptional sintering resistance. Journal of the European Ceramic Society, 41(13), 6670-6676.

[102] Xu, Z., He, L., Zhong, X., Mu, R., He, S. & Cao, X. (2009). Thermal barrier coating of lanthanum-zirconium-cerium composite oxide made by electron beam-physical vapor deposition. Journal of Alloys and Compounds, 478, 168-172.

[103] Drebushchak, V.A. (2020). Thermal expansion of solids: Review on theories. Journal of Thermal Analysis and Calorimetry, 142(2), 1097-1113.

[104] Feng, D., Dong, Y., Zhang, L., Ge, X., Zhang, W., Dai, S. & Qiao, Z. (2020). Holey lamellar high-entropy oxide as an ultra-high-activity heterogeneous catalyst for solvent-free aerobic oxidation of benzyl alcohol. Angewandte Chemie International Edition, 59, 19503-19509.

[105] Deng, C., Wu, P., Zhu, L., He, J., Tao, D.J., Lu, L. et al. (2020). High-entropy oxide stabilized molybdenum oxide via high temperature for deep oxidative desulfurization. Applied Materials Today, 20, 100680.

[106] Sun, Y. & Dai, S. (2021). High-entropy materials for catalysis: A new frontier. Science Advances, 7(20), eabg1600.

[107] Yeh, J.-W. (2013). Alloy design strategies and future trends in high-entropy alloys. JOM, 65, 1759-1771.

[108] Khorshidi, A., Violet, J., Hashemi, J. & Peterson, A.A. (2018). How strain can break the scaling relations of catalysis. Nature Catalysis, 1, 263-268.

[109] Tsai, K.-Y., Tsai, M.-H. & Yeh, J.-W. (2013). Sluggish diffusion in Co-Cr-Fe-Mn-Ni high-entropy alloys. Acta Materialia, 61, 4887-4897.

[110] Zhang, Z., Yang, S., Hu, X., Xu, H., Peng, H., Liu, M. et al. (2019). Mechanochemical nonhydrolytic sol-gel strategy for the production of mesoporous multimetallic oxides. Chemistry of Materials, 31, 5529-5536.

[111] Shu, Y., Bao, J., Yang, S., Duan, X. & Zhang, P. (2021). Entropy-stabilized metal-CeOx solid solutions for catalytic combustion of volatile organic compounds. American Institute of Chemical Engineers Journal, 67, e17046.

[112] Xu, H., Zhang, Z., Liu, J., Do-Thanh, C.-L., Chen, H., Xu, S., et al. (2020). Entropy-stabilized single-atom Pd catalysts via high-entropy fluorite oxide supports. Nature Communications, 11, 3908.

Perovskite-Structured Oxides and High-Entropy Perovskites

Perovskite, a mineral with the chemical formula $CaTiO_3$, was discovered in 1839 by a Prussian mineralogist, Gustav Rose, within the mineral deposits of the Ural Mountains. Named in honour of Russian mineralogist Count Lev Aleksevich von Petrovski, natural perovskite crystals exhibit a Mohs hardness of 5.5-6 and a density ranging from 4000 to 4300 kg/m³ [1]. Initially believed to be cubic, the crystal structure of natural perovskite crystals was later determined to be orthorhombic.

The mineral Perovskite has lent its name to a broad class of compounds known as perovskites, which are characterized by a general formula similar to (or derived from) ABX_3. To date, a huge number of compounds adopting the perovskite structure have been identified. The extensive range of phases within this family can be explained by considering perovskites as simple ionic compounds. Typically, A represents a large cation, B denotes a medium-sized cation, and X refers to an anion (mostly oxygen). Figure 1 shows a generic ABX_3 perovskite structure.

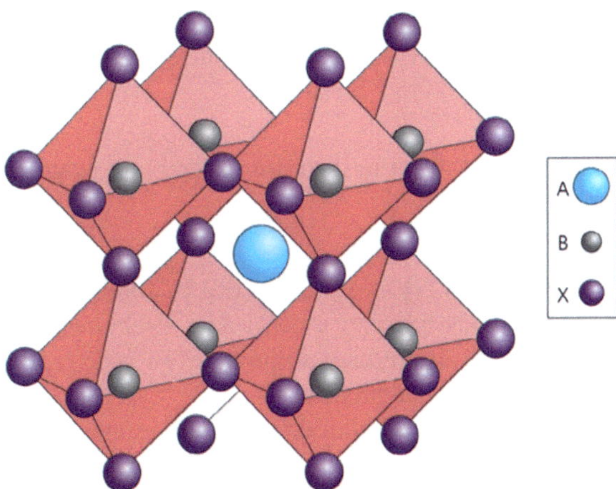

Fig. 1: A generic ABX_3 perovskite structure.

The perovskite family boasts an array of properties and potential applications across various technological sectors. The structural flexibility of perovskite compounds, attributed to the vast number of combinations for A, B, and X elements, has garnered significant interest from academic and industrial communities. Some notable properties and applications of perovskite compounds include superconductivity [2, 3], ferroelectricity [4, 5], photovoltaics [6, 7], catalysis [8-12], and optoelectronics [13,14]. However, despite their promising potential, several challenges, such as stability and toxicity, must be addressed before perovskites can be widely adopted in various industries [15-17].

The industrial significance of perovskites emerged when researchers discovered the valuable dielectric and ferroelectric properties of barium titanate ($BaTiO_3$) in the 1940s [18]. This material was quickly utilized in electronics for applications such as capacitors and transducers. In the subsequent decades, researchers aimed to enhance the properties of $BaTiO_3$ by investigating the relationship between structure and properties in a large variety of ceramic perovskite-related phases with general compositions of ABO_3. As a result, a vast array of new phases were synthesized [19-21].

The ideal perovskite structure, also known as the "aristotype" perovskite structure, is cubic and is exhibited by strontium titanate ($SrTiO_3$) at room temperature (though not at all temperatures) [22]. From a crystallographic standpoint, the ideal perovskite structure is rigid, as the unit cell has no adjustable atomic position parameters. Consequently, any changes in composition must be accommodated by a change in lattice parameter, which is essentially the sum of the anion-cation bond lengths.

In the cubic unit cell, the edge length 'a' is equivalent to twice the B-X bond length, so also in the ideal perovskite structure, the cubic unit cell edge 'a' is equal to twice the B-X bond length:

$$2(B-X) = a \tag{1}$$

The width of the cuboctahedral cage site, which is equal to the square root of 2 times 'a', corresponds to twice the A-X bond length:

$$2(A-X) = \sqrt{2}a \tag{2}$$

For the ideal structure to form, the ratio of the bond lengths should be:

$$(A-X)/(B-X) = \sqrt{2} \tag{3}$$

In 1926, Goldschmidt [23] first utilized this relationship (3) to predict the likelihood of a pair of ions forming a perovskite structure phase. Only a few crystal structures had been determined at the time, so ionic radii were used as a substitute for measured bond lengths.

Therefore, according to Goldschmidt's rule, it was assumed that the cations just "touch" the surrounding anions to form a stable structure. This led to the following equation for the tolerance factor (t):

$$t = \left(r_a + r_x \right) / \sqrt{2} \left(r_b + r_x \right) = 1 \tag{4}$$

In equation (4), r_a, r_b, and r_x represent the ionic radii of the A, B, and X ions, respectively. The tolerance factor (t) can help predict the stability and formation of perovskite structures, providing insights into potential material compositions that may adopt this crystal structure.

The tolerance factor (t) is a very useful parameter to predict the formation of perovskite structure phases. In particular, Goldschmidt suggested that a cubic perovskite structure phase would form if the tolerance factor (t) is close to 1.0.

Using ionic radii that correspond to the ions' coordination geometry is crucial. Therefore, r_a should be suitable for 12 coordination, r_b for octahedral coordination, and r_x for linear coordination. Despite its simplicity, the Goldschmidt tolerance factor has proven to have great predictive power, especially for oxides where ionic radii are determined with the highest accuracy. For the ideal perovskite structure, t should be equal to 1.0. Still, it has been found empirically that if t falls within the approximate range of 0.9-1.0, a cubic perovskite structure is the expected outcome.

If $t > 1$, meaning large A and small B, a hexagonal packing of AX_3 layers is preferred, forming hexagonal phases such as $BaNiO_3$ type [1]. When t is approximately 0.71-0.9, the structure, particularly the octahedral framework, distorts to close the cuboctahedral coordination polyhedron, leading to a crystal structure with lower symmetry than cubic. For even lower values of t, the A and B cations are similar in size and are associated with structures like ilmenite ($FeTiO_3$) or C-type rare earth structures like Ln_2O_3 [1].

Table 1 reports some of the most representative perovskite compounds, indicating their crystal structure and unit cell parameters.

Diffusion and Ionic Conductivity in Perovskites

In perovskites, native point defect populations are usually sufficient for low diffusivity. However, for fast diffusion of cations and anions, either a significant population of vacancies on the relevant sublattices or fairly open regions, such as those found in layered phases, are necessary. To increase diffusivity, perovskites with near-cubic structures are usually doped to create vacancies on the appropriate sublattice.

Cation diffusion is generally described by an Arrhenius-type relationship:

$$D = D_0 * exp\left(-E / RT \right) \tag{5}$$

Where D represents diffusivity, D_0 denotes the pre-exponential factor, E represents the activation energy for diffusion, and T stands for the temperature in Kelvin (K).

Table 1: Most representative perovskite systems

Composition	Crystal structure	Unit cell		
		a (nm)	b (nm)	c (nm)
$AgMgF_3$	Cubic	0.41162		
$CsPbI_3$	Cubic	0.62894		
$KTaO_3$	Cubic	0.40316		
$SrTiO_3$	Cubic	0.3905		
$BaSnO_3$	Cubic	0.4117		
$SrCoO_3$	Cubic	0.3855		
$SrMoO_3$	Cubic	0.39761		
$BaCeO_3$	Cubic	0.0470		
$KCuF_3$	Tetragonal	0.56086		0.76281
$BaTiO_3$	Tetragonal	0.39906		0.40278
$PbTiO_3$	Tetragonal	0.3902		0.4143
$NaMgF_3$	Orthorombic	0.48904	0.52022	0.71403
$KNbO_3$	Orthorombic	0.3971	0.5697	0.5723
$CaTiO_3$	Orthorombic	0.54035	0.54878	0.76626
$PbZrO_3$	Orthorombic	0.58822	1.17813	0.82293
$GdFeO_3$	Orthorombic	0.5349	0.5609	0.7669
$LaTiO_3$	Orthorombic	0.563676	0.561871	0.791615
$LaCrO_3$	Orthorombic	0.55163	0.54790	0.77616
$LaFeO_3$	Orthorombic	0.55563	0.55630	0.78535
$SrZrO_3$	Orthorombic	0.57093	0.57053	0.80676
$LaMnO_3$	Orthorombic	0.55397	0.54891	0.77928
$CeMnO_3$	Orthorombic	0.5537	0.5557	0.7821
$BaCeO_3$	Orthorombic	0.6250	0.6260	0.8810
$LaCoO_3$	Trigonal	0.54437	0.54437	1.30957
$LaNiO_3$	Trigonal	0.54524	0.54524	1.31572
$LiNbO_3$	Trigonal	0.51381	0.51381	1.38481
$LiTaO_3$	Trigonal	0.51528	0.51528	1.37795
$BaNiO_3$	Hexagonal	0.5629		0.4811
$SrCoO_3$	Hexagonal	0.5486		0.420
$BaCoO_3$	Hexagonal	0.56525		0.47629
$BaMnO_3$	Hexagonal	0.56991		0.48148
$KNiCl_3$	Hexagonal	1.1795		0.5926

The activation energy E can be described as the energy needed to move an ion from one stable position, typically through a bottleneck of surrounding ions, to a neighbouring stable position.

Similar to diffusion, ionic conductivity of cations and anions in perovskites requires either open regions in the structure or a significant population of vacancies on the appropriate sublattice to facilitate ionic movement. Substitution is commonly used to create vacancies in perovskites with near-cubic structures, thereby increasing their ionic conductivity.

The absence of cations with variable valence is necessary for strict ionic conductivity. If variable valence cations are present inside a perovskite compound, electronic conductivity may also occur, generally dominating ionic conductivity in magnitude.

Ionic conductivity can be described by an Arrhenius-type equation too, often written as:

$$\sigma = \sigma_0 * exp\left(-E \: / RT\right) \qquad (6)$$

In Equation (6), σ represents ionic conductivity, T is the temperature expressed in Kelvin (K), σ_0 represents the pre-exponential factor, and E is the activation energy for ionic conduction. A high concentration of vacancies leads to a significant increase in oxide ion conductivity.

Oxide ion conductivity (and diffusion) occurs via a mechanism where the interstitial ion moves into a normally occupied site while simultaneously displacing the resident ion into an adjacent interstitial position.

Proton Conductivity in Perovskites

Perovskite oxides with proton-conducting properties are currently the subject of extensive research due to their potential applications in electrochemical devices such as fuel cells, electrochromic displays, and hydrogen sensors [24-28].

Perovskite oxides can be transformed into efficient proton conductors through point defect chemistry. Substituted (2,4) $A^{2+}B^{4+}O_3$ perovskites, such as $BaTiO_3$, $BaCeO_3$, and $BaZrO_3$, as well as the $A_2B_2O_5$ perovskite derivative $Ba_2In_2O_5$, are among the most studied materials as proton conductors [29-31].

$Ba_2In_2O_5$ adopts the brownmillerite structure at lower temperatures and begins to disorder around 900°C, forming a cubic perovskite phase with a high concentration of oxygen vacancies at higher temperatures. Proton conduction is achieved by incorporating water molecules, which react directly with the oxide at elevated temperatures, forming OH^- ions and H^+ ions (protons). These ions occupy oxygen vacancies and form defects within the structure. The weaker hydrogen bonding in the material is believed to be the reason for its high proton conductivity [1].

$BaCeO_3$ and $BaZrO_3$ perovskites, which are insulating solids when prepared in air, can be converted into oxygen-deficient phases by doping the B-sites with fixed-valence trivalent M^{3+} ions [32], such as Gd^{3+} or Y^{3+}. Charge balance is maintained by introducing oxygen vacancies, and the formation of one oxygen vacancy balances the presence of two M^{3+} substituents.

Proton conductivity in these kinds of materials is also induced by a reaction with water vapor, similar to $Ba_2In_2O_5$. At higher temperatures, reaction with water vapor causes disproportionation of water molecules at the oxide surface, with the resultant OH^- ions occupying oxygen vacancies in the structure and protons hydrogen bonded to existing lattice oxygen.

In all proton-conducting perovskites, the hydration process can be formally described by a defect equation:

$$H_2O(g) + O_o + V_o^{\cdot\cdot} \rightarrow 2OH_o^{\cdot} \tag{7}$$

The equilibrium constant (K) for reaction (7) is:

$$K = \left[OH_o^{\cdot}\right]^2 / (\left[V_o^{\cdot\cdot}\right] * \left[O_o\right] * p_{H_2O} \tag{8}$$

The number of mobile protons available for conductivity increases proportionally with the partial pressure of water vapor, as described by:

$$\left[OH_o^{\cdot}\right]^2 = K * [V_o^{\cdot\cdot}] * [O_o] * P_{H_2O} \tag{9}$$

This relationship results in a parabolic hydration isotherm with a curve dependent on the dopant concentration. The values of $[V_o^{\cdot\cdot}]$ and the equilibrium constant K can be deduced from the curve. The relationship between water vapor pressure and conductivity can be harnessed experimentally for humidity sensors. The reaction is also sensitive to temperature, with higher water uptake and increased proton conduction achieved at lower temperatures.

Determining the true stoichiometry of proton-conducting perovskite compounds and the proton positions within them is challenging due to traditional diffraction techniques averaging over large volumes of structure. This makes it difficult to precisely understand the detailed mechanism of conduction in such perovskites.

Dielectric Properties in Perovskites

Perovskites play a crucial role in various electronic applications, ranging from basic capacitors to dielectric resonators employed in mobile phones, satellite communications, TV broadcasting, and more [33, 34].

The dielectric properties of bulk perovskites stem from the presence of polarizable components in the crystal, including cation displacements, octahedral tilting and distortions, grain boundaries, and point defects. Relative permittivity (ε_r) is the fundamental parameter describing a dielectric in a static electric field, while the complex relative permittivity $(\varepsilon_r' + i\,\varepsilon_r'')$ is a function of the frequency of the applied electric field in varying electric fields. Another important metric for dielectric performance in varying electric fields is energy loss, typically expressed as the loss tangent (tan δ), which is equal to $\varepsilon_r''/\varepsilon_r'$. Relative permittivity tends to increase with temperature as polarizable constituents become more mobile and declines as the frequency of the applied electric field increases.

However, the dielectric properties of a simple perovskite are seldom ideal for a specific application. To improve perovskite dielectric materials, a common approach is to start with a material with desirable qualities (e.g., $CaTiO_3$, $BaTiO_3$, or $ZrTiO_3$) and then substitute on the A- and B-sites to modify the properties accordingly [1]. The resulting dielectric response depends on numerous factors, and small amounts of dopant can lead to significant changes in both relative permittivity and loss tangent.

The relationship between preparation conditions, microstructure, and properties is clearly demonstrated by ceramic oxides referred to as colossal dielectric constant (CDC) materials [35]. The most notable of these is the cubic perovskite phase $CaCu_3Ti_4O_{12}$ (CCTO), along with closely related materials such as $SrCu_3Ti_4O_{12}$, and $Na_{0.5}Bi_{0.5}Cu_3Ti_4O_{12}$ [36-38]. All these materials adopt a cubic perovskite structure.

The relative permittivity of such compounds is around 10^5 at low frequencies, remaining nearly constant across frequency and temperature over a wide temperature range of 100-500 K. These phases are ideal for use in miniaturized electronic components and energy storage applications due to their exceptional dielectric properties [39].

Magnetic Properties in Perovskites

Magnetic materials can be approached from the perspective of free electron band theory or a modification of the familiar ionic model. Some perovskites can be better understood in terms of free electron theory, forming a group of metallic phases analogous to metallic elemental magnets. In these cases, electron spin interactions, typically significant at lower temperatures, may be strong enough to result in parallel (ferromagnetic) alignment. At higher temperatures, these materials transform into metal-like Pauli paramagnetic compounds, displaying almost no temperature dependence for the magnetic susceptibility.

However, many perovskites are magnetic due to the incorporation of paramagnetic cations within the perovskite structure, meaning cations that display a magnetic moment themselves. These materials can often be considered in ionic terms. The most important magnetic species are transition metal and lanthanoid cations, which have incompletely filled d and f shells. The geometry of the crystallographic site occupied by these cations is crucial, as it governs the d and f cation orbital energy level positions resulting from crystal-field/ligand-field interactions.

In particular, for the $3d$ transition metal cations, the magnetic moment mainly arises from the electron spin alone and is often referred to simply as the spin (S). Depending on the strength of the crystal-field interaction, the total spin can often take one of several values: high spin (HS), intermediate spin (IS), or low spin (LS). The magnetic behavior is then associated with the organization between the magnetic moments or spin on these cations. At higher temperatures, the magnetic moments are disordered, leading to paramagnetic compounds with a temperature-dependent magnetic susceptibility obeying the Curie or Curie–Weiss laws [40]. At lower temperatures, magnetic moments can interact, resulting in antiparallel (antiferromagnetic) or parallel (ferromagnetic) alignment, as well as various other magnetic configurations.

In perovskites containing paramagnetic cations, the magnetic moments in the structure are arranged randomly at higher temperatures. This arrangement is dynamic, with the orientation of the dipoles continually changing due to thermal effects, resulting in the paramagnetic state. When exposed to a magnetic field, these magnetic dipoles will attempt to orient themselves parallel to the magnetic flux density in the solid. However, thermal agitation opposes this alignment. As a result, the paramagnetic susceptibility (χ) varies with temperature. For a simple paramagnetic system, this dependence follows the Curie law:

$$X = C/T \tag{10}$$

Where C is the Curie constant, and T is the temperature expressed in Kelvin (K).

Ferromagnetic perovskites undergo a transformation to the paramagnetic state above the Curie temperature (T_C). Above this temperature, thermal energy is sufficient to disrupt the ordered arrangement of magnetic moments, causing them to become disordered and resulting in the loss of the ferromagnetic properties.

In cubic perovskite-related structures, the B-site cations are situated in apex-shared octahedra, while the A-site cations are located in the cage sites between the octahedra. The cation-anion-cation geometry is linear, which, according to the superexchange mechanism [41], suggests that these phases should favor antiferromagnetic (antiparallel) alignment of the magnetic moments.

Schematically, a cation with an overall spin-up configuration can interact with a spin-down electron in a filled oxygen p orbital. Consequently, the other (spin-down) electron in the oxygen p orbital must then induce a favorable exchange to enforce an overall spin-down configuration in the other cation. This result is generally true, particularly for perovskites with a single magnetic cation, such as $LaFeO_3$ [42]. These compounds are typically insulators or high-resistivity semiconductors.

Thermal Properties in Perovskites

Thermal expansion is a crucial physical property for perovskites used in high-temperature applications, such as solid oxide fuel cells operating above 800 °C [43]. Mismatch in the thermal expansion of cell components (cathode, electrolyte, and anode) can lead to early cell failure. The magnitude of thermal expansion in many perovskites is rooted in the thermal behavior of the BX_6 octahedra and is associated with octahedral tilt, distortion, and the bonding between the B-cation and surrounding anions [1]. These factors are all susceptible to modification as the temperature rises and can contribute to anomalies in thermal expansion characteristics.

Thermal expansion of perovskites may also depend on features such as magnetic and spin ordering. If a perovskite transitions from a magnetically ordered phase, like antiferromagnetic, to paramagnetic, the thermal expansion will typically show an anomaly [44]. This has been reported in the well-studied lanthanoid titanates, $LnTiO_3$ (Ln is a generic lanthanide), which exhibit both antiferromagnetic and ferromagnetic regimes [45]. A similar situation occurs in $LaCoO_3$ [46].

Negative thermal expansion (NTE) occurs in some perovskite compounds, which contract as the temperature increases over a specific temperature range. Thermal contraction is not the result of a single mechanism, and there are several factors contributing to this behavior in different perovskite structures.

One example of NTE is found in $PbTiO_3$ [47]. At lower temperatures, $PbTiO_3$ has a tetragonal unit cell due to the displacement of Ti^{4+} cations away from the centers of the TiO_6 octahedra along the c-axis. As the temperature increases, the distortion decreases due to changes in vibrational energy and a decrease in anion-anion repulsion. This leads to a shortening of the long diagonal and a decrease in the c-axis. Simultaneously, the undistorted diagonals expand, causing the a-axis and b-axis to increase normally. The contraction outweighs the expansion, resulting in an overall thermal contraction up to the Curie temperature of 490 °C. The mean thermal expansion coefficient (α) for $PbTiO_3$ ceramics is approximately -1.99×10^{-5} K^{-1}.

Above the transition temperature, $PbTiO_3$ transforms to the ideal cubic structure, and normal thermal expansion is observed as the temperature increases, with a thermal expansion coefficient of approximately 3.72×10^{-5} K^{-1}.

Generally speaking, negative thermal expansion in perovskites could arise from the interplay between various factors, such as changes in vibrational energy, anion-anion repulsion, and structural distortions. These effects can lead to an overall contraction of the material over a certain temperature range, making such perovskites very interesting for specific technological application.

High-Entropy Perovskites

Following the pioneering work on rocksalt-structured high-entropy oxides, Jiang et al. [48] demonstrated that entropy stabilization could be applied to more complex oxide systems, specifically perovskite oxides (ABO_3), in which cations share two or more Wyckoff sites.

Normally in perovskites, the addition of more cations in higher amounts than typical small doping ranges lead to a strong tendency to form secondary phases. This is because complex oxides like perovskites have a higher chemical complexity compared to simpler oxide structures like the rocksalt one, and the formation of alternative structures of the type $A_xB_yX_z$ (e.g. Ruddlesden-Popper type phases) is often favored in such more complex systems [49].

Jiang et al. substituted the B-site with five different elements, while keeping the A-site fixed to one or at most two elements. Particularly, they have successfully synthesized six homogeneous single-phase high-entropy ABO_3 perovskite oxides, suggesting that having a Goldschmidt tolerance factor close to unity ($t \approx 1.00$) is a necessary but not sufficient criterion for forming a single high-entropy perovskite phase. They also found that single solid-solution phases are formed in Sr-based compositions with $t < 1$ and in Ba-based compositions with $t > 1$.

The work of Jiang et al. represents the first successful synthesis of high-entropy perovskite oxides, which are single solid-solution phases of multi-cation perovskite oxides with high configurational entropies of ≥ 1.5 R per mole (R is the universal gas constant).

Their pioneering investigation highlighted the potential for entropy stabilization in more complex oxide systems, opening up new possibilities for the understanding and control of entropy stabilization in perovskite systems leading to the development of novel materials with tailored properties for the huge possible applications of perovskites (mainly energy storage, electronics, and catalysis).

Following Jiang et al., Sarkar et al. [50] studied eleven different high-entropy perovskites, demonstrating the stabilization in a single-phase for six of them, including the extremely complex 10-cationic system $(Gd_{0.2}La_{0.2}Nd_{0.2}Sm_{0.2}Y_{0.2})$ $(Co_{0.2}Cr_{0.2}Fe_{0.2}Mn_{0.2}Ni_{0.2})O_3$. Moreover, Sarkar et al.'s study led to the following conclusions:

- Unfavorable enthalpic interactions in some of the lower entropy (heptanary) systems might not be overcome by the entropic contribution;
- The entropy-driven reversible transformation from a multiphase to single-phase upon cyclic heat treatment in perovskite systems has been observed for the $(Gd_{0.2}La_{0.2}Nd_{0.2}Sm_{0.2}Y_{0.2})MnO_3$ system, but not for other single-phase systems;
- Large metric distortion from cubic symmetry, along with strong tilting of the BO_6 polyhedra and lowering of the effective coordination number of A-site cations, is observed for systems with smaller A-site cations and/or larger B-site cations;
- Enthalpic factors (such as cation-size-induced metric distortion, the affinity of certain elements towards a specific oxidation state, and Jahn-Teller distortion) favored an ordered scenario, potentially making the formation of the cubic "aristotype" structure in high-entropy perovskites energetically expensive.

Sarkar et al. conclusions helped deepen the understanding of the complex interplay between entropy, enthalpy, and the structural stability of high entropy perovskite materials, paving the way for further exploration and development of advanced materials with unique properties and performance characteristics.

In 2021, Ma et al. [51] acknowledged that most high-entropy perovskite oxides synthesized so far [48, 50] are composed of cations of the same valence, limiting the choice of cations and focusing primarily on equimolar compounds. This focus neglects non-equimolar compounds, which could potentially lead to high-performance materials. In fact, to fully exploit the tunability of perovskite oxides, non-equimolar compounds with cations of different valences need to be explored.

In this regard, Ma et al. have proposed a strategy for designing high-entropy oxides called the valence-combination strategy. This strategy consists of two steps to select suitable cations for the synthesis of high entropy perovskites with cations of different valences:

- In the first step, possible valence combinations for a given material system are designed by considering electric balance. Each valence combination is called a subsystem;
- In the second step, cations with proper size are selected for each subsystem by considering geometrical factors, such as Pauling's first rule and Goldschmidt tolerance factor.

The valence of the B-sites in perovskite oxides can be $+3$, $+4$, or $+5$, with corresponding A-site valences of $+3$, $+2$, and $+1$, respectively. Considering that monovalent cations usually have a large radius and are more likely to enter the A-site, and that there are very few cations with a valence larger than 6, only cations with valences ranging from $+2$ to $+6$ have been considered. Based on these criteria, when the average valence of B-site cations is $+3$, $+4$, and $+5$, the number of valence combinations (subsystems) is 66, 86, and 66, respectively.

For each valence combination, there are a huge number of possible compositions (ratios of substitutional cations). However, only one composition has the maximum configurational entropy. To find such composition, it is possible to vary the composition in 1% intervals and then calculate the entropy of each composition.

The number of subsystems with the maximum configurational entropy greater than 1.5 R (R is the universal gas constant) is 29, 56, and 29 for average B-site cation valences of $+3$, $+4$, and $+5$, respectively. The number of valence combinations with a configurational entropy of 1.61 R (i.e. the highest possible configurational entropy for five substitutional high-entropy materials) is 6, 12, and 6 for average B-site cation valences of $+3$, $+4$, and $+5$, respectively.

By considering the maximum configurational entropy, it is possible to strategically select compositions that are more likely to yield high-performance materials with unique properties.

By applying this strategy to equimolar B-site substitutional $A(5B_{0.2})O_3$ perovskite, Tang et al. selected eight combinations with the maximum configurational entropy and successfully synthesized (out of these eight combinations) six single-phase perovskite oxides, whose exact composition is reported in the following Table 2.

                 Different Classes of High-Entropy Materials

Table 2: High-entropy perovskites synthesized via the valence combination design method proposed in [51]

Composition	Valence combination	Single-phase?
$Ba(Mg_{0.17}Zn_{0.17}Ti_{0.21}Nb_{0.22}W_{0.23})O_3$	+2 +2 +4 +5 +6	Yes
$Ba(Mg_{0.15}Yb_{0.17}Ti_{0.21}Nb_{0.23}Ta_{0.24})O_3$	+2 +3 +4 +5 +5	Yes
$Ba(Mg_{0.16}Y_{0.19}Yb_{0.19}Nb_{0.22}W_{0.24})O_3$	+2 +3 +3 +5 +6	Yes
$Ba(Mg_{0.22}Zn_{0.23}Nb_{0.20}W_{0.17}Mo_{0.18})O_3$	+2 +2 +5 +6 +6	No
$Ba(Mg_{0.23}Yb_{0.22}Ti_{0.21}W_{0.17}Mo_{0.17})O_3$	+2 +3 +4 +6 +6	Yes
$Ba(Mg_{0.24}Yb_{0.22}Nb_{0.19}Ta_{0.19}W_{0.16})O_3$	+2 +3 +5 +5 +6	No
$Ba(Mg_{0.25}Ti_{0.21}Zr_{0.20}Nb_{0.18}W_{0.16})O_3$	+2 +4 +4 +5 +6	Yes
$Ba(Y_{0.22}Yb_{0.22}Dy_{0.22}W_{0.17}Mo_{0.17})O_3$	+3 +3 +3 +6 +6	Yes

It has also been observed (like in the case of [48]) that the formation of single-phase high-entropy perovskite oxides does not have a clear and simple relationship with the Goldschmidt structural tolerance factor (t), suggesting that the formation of ordered perovskite structures can be correlated to the valence-difference factor.

The study of Ma et al. also proposes the possibility of applying the valence-combination strategy in the design of non-stoichiometric materials, allowing for a more systematic and efficient exploration of a wider range of high-entropy perovskite compositions. In fact, by extending this strategy to non-stoichiometric materials, researchers can further explore and exploit the unique properties and tunability of perovskite oxides, opening up new possibilities for the development of high-performance materials with a wide range of applications.

In terms of high-entropy perovskites design strategies, Spiridigliozzi et al. [52] successfully applied the cluster-plus-atom design model [53, 54] to successfully synthesize a novel high-entropy perovskite oxide, $Ba(Ce_{0.2}Zr_{0.2}Gd_{0.2}Y_{0.2}La_{0.2})O_{2.7}$, derivative of the proton-conducting barium cerate/zirconates, and demonstrate its reversible entropy-driven single-phase stabilization at around 1300 °C.

Furthermore, while the stable crystal structure of $Ba(Ce_{0.2}Zr_{0.2}Gd_{0.2}Y_{0.2}La_{0.2})O_{2.7}$ could be predicted being cubic using the Goldschmidt tolerance factor, such equilibrium structure depends more on diffusion kinetics compared to conventional perovskites (confirming the finding of Sarkar et al. [50] about the cubic "aristotype" structure in high-entropy perovskites being energetically expensive). In fact, two high-entropy perovskite oxides can be derived from the designed composition ($Ba(Ce_{0.2}Zr_{0.2}Gd_{0.2}Y_{0.2}La_{0.2})O_{2.7}$), i.e. an orthorhombic phase and a cubic phase, with the orthorhombic phase initially obtained at 1300 °C and the cubic phase (predicted by the Goldschmidt's factor) only formed at 1550 °C.

From the point of view of their properties, up to now, many high-entropy perovskite systems have been proposed in literature to be possible game changers in various technological sectors, such as energy conversion and storage, catalysis and electrocatalysis, fuel cells and solar cells [55].

A notable example of possible technological breakthroughs favored by high-entropy perovskites is the high-entropy perovskite oxides $[(Bi,Na)_{0.2}(La,Li)_{0.2}(Ce,K)_{0.2}Ca_{0.2}Sr_{0.2}]TiO_3$ proposed by Yan et al. [56] and applied as an anode material for lithium-ion batteries (LIBs)

Such high-entropy perovskite anode demonstrated outstanding cycle capacity and superior capacity retention (approximately 100% at 1000 mAh g^{-1} after 300 cycles), such excellent performance being attributable to the entropy-stabilized structure and the charge compensation mechanism present in it.

References

[1] Tilley, R.J. (2016). Perovskites: Structure-Property Relationships. John Wiley & Sons.

[2] Schneemeyer, L.F., Waszczak, J.V., Zahorak, S.M., van Dover, R.B. & Siegrist, T. (1987). Superconductivity in rare earth cuprate perovskites. Materials Research Bulletin, 22(11), 1467-1473.

[3] Liu, S.Y., Meng, Y., Liu, S., Li, D.J., Li, Y., Liu, Y., Shen, Y. & Wang, S. (2017). Phase stability, electronic structures, and superconductivity properties of the $BaPb_{1-x}Bi_xO_3$ and $Ba_{1-x}K_xBiO_3$ perovskites. Journal of the American Ceramic Society, 100(3), 1221-1230.

[4] Rørvik, P.M., Grande, T. & Einarsrud, M.A. (2011). One-dimensional nanostructures of ferroelectric perovskites. Advanced Materials, 23(35), 4007-4034.

[5] Chu, M.W., Szafraniak, I., Scholz, R., Harnagea, C., Hesse, D., Alexe, M. & Gösele, U. (2004). Impact of misfit dislocations on the polarization instability of epitaxial nanostructured ferroelectric perovskites. Nature Materials, 3(2), 87-90.

[6] Park, N.G. (2015). Perovskite solar cells: An emerging photovoltaic technology. Materials Today, 18(2), 65-72.

[7] Huang, J., Yuan, Y., Shao, Y. & Yan, Y. (2017). Understanding the physical properties of hybrid perovskites for photovoltaic applications. Nature Reviews Materials, 2(7), 1-19.

[8] Hwang, J., Rao, R.R., Giordano, L., Katayama, Y., Yu, Y. & Shao-Horn, Y. (2017). Perovskites in catalysis and electrocatalysis. Science, 358(6364), 751–756.

[9] Xu, X., Zhong, Y. & Shao, Z. (2019). Double perovskites in catalysis, electrocatalysis, and photo (electro) catalysis. Trends in Chemistry, 1(4), 410-424.

[10] Polo-Garzon, F. & Wu, Z. (2018). Acid–base catalysis over perovskites: A review. Journal of Materials Chemistry A, 6(7), 2877–2894.

[11] Arandiyan, H., Mofarah, S.S., Sorrell, C.C., Doustkhah, E., Sajjadi, B., Hao, D., Wang, Y., Sun, H., Ni, B., Rezaei, M., Shao, Z. & Maschmeyer, T. (2021). Defect engineering of oxide perovskites for catalysis and energy storage: Synthesis of chemistry and materials science. Chemical Society Reviews, 50(18), 10116–10211.

[12] Li, M., Han, N., Zhang, X., Wang, S., Jiang, M., Bokhari, A. et al. (2022). Perovskite oxide for emerging photo (electro) catalysis in energy and environment. Environmental Research, 205, 112544.

[13] Dong, Y., Zhang, Y., Li, X., Feng, Y., Zhang, H. & Xu, J. (2019). Chiral perovskites: Promising materials toward next-generation optoelectronics. Small, 15(39), 1902237.

[14] Zhang, Y., Lim, C.K., Dai, Z., Yu, G., Haus, J.W., Zhang, H. & Prasad, P.N. (2019). Photonics and optoelectronics using nano-structured hybrid perovskite media and their optical cavities. Physics Reports, 795, 1–51.

[15] Ren, M., Qian, X., Chen, Y., Wang, T. & Zhao, Y. (2022). Potential lead toxicity and leakage issues on lead halide perovskite photovoltaics. Journal of Hazardous Materials, 426, 127848.

[16] Zhu, T., Yang, Y. & Gong, X. (2020). Recent advancements and challenges for low-toxicity perovskite materials. ACS Applied Materials & Interfaces, 12(24), 26776-26811.

[17] Román-Vázquez, M., Vidyasagar, C.C., Muñoz-Flores, B.M. & Jiménez-Pérez, V.M. (2020). Recent advances on synthesis and applications of lead-and tin-free perovskites. Journal of Alloys and Compounds, 835, 155112.

[18] Kay, H.F. & Vousden, P. (1949). XCV. Symmetry changes in barium titanate at low temperatures and their relation to its ferroelectric properties. The London, Edinburgh, and Dublin Philosophical Magazine and Journal of Science, 40(309), 1019–1040.

[19] Zhitomirskii, I.D., Burdina, K.P., Chechernikova, O.I., Gagulin, V.V., Sevast'yanova, L.G. & Venevtsev, Y.N. (1986). Dielectric and magnetic properties of new perovskites PbMn/SUB 2/3/B/SUB 1/3/O/sub 3/(B = Mo, Te, Re). Inorganic Materials (Engl. Transl.); (United States), 21(8).

[20] Bednorz, J.G. & Müller, K.A. (1988). Perovskite-type oxides—The new approach to high-Tc superconductivity. Reviews of Modern Physics, 60(3), 585.

[21] Bhalla, A.S., Guo, R. & Roy, R. (2000). The perovskite structure—A review of its role in ceramic science and technology. Materials Research Innovations, 4(1), 3–26.

[22] Yamanaka, T., Ahart, M., Mao, H.K. & Yan, H. (2018). New high-pressure tetragonal polymorphs of SrTiO—Molecular orbital and Raman band change under pressure. Journal of Physics: Condensed Matter, 30(26), 265401.

[23] Goldschmidt, V.M. (1926). Die Gesetze der Krystallochemie. Naturwissenschaften, 14, 477–485.

[24] Zając, W., Rusinek, D., Zheng, K. & Molenda, J. (2013). Applicability of Gd-doped BaZrO₃, SrZrO₃, BaCeO₃ and SrCeO₃ proton conducting perovskites as electrolytes for solid oxide fuel cells. Open Chemistry, 11(4), 471–484.

[25] Granqvist, C.G. (2019). Electrochromism and Electrochromic Devices. The CRC Handbook of Solid State Electrochemistry, 587–615.

[26] Hossain, M.K., Chanda, R., El-Denglawey, A., Emrose, T., Rahman, M.T., Biswas, M.C. & Hashizume, K. (2021). Recent progress in barium zirconate proton conductors for electrochemical hydrogen device applications: A review. Ceramics International, 47(17), 23725–23748.

[27] Tan, X., Shen, Z., Bokhari, A., Qyyum, M.A. & Han, N. (2022). Insights on perovskite-type proton conductive membranes for hydrogen permeation. International Journal of Hydrogen Energy, 48(68), 26541–26550.

[28] Li, S. & Irvine, J.T. (2021). Non-stoichiometry, structure and properties of proton-conducting perovskite oxides. Solid State Ionics, 361, 115571.

[29] Hossain, S., Abdalla, A.M., Jamain, S.N.B., Zaini, J.H. & Azad, A.K. (2017). A review on proton conducting electrolytes for clean energy and intermediate temperature-solid oxide fuel cells. Renewable and Sustainable Energy Reviews, 79, 750–764.

[30] Kochetova, N., Animitsa, I., Medvedev, D., Demin, A. & Tsiakaras, P. (2016).

Recent activity in the development of proton-conducting oxides for high-temperature applications. RSC Advances, 6(77), 73222–73268.

[31] Jankovic, J., Wilkinson, D.P. & Hui, R. (2010). Proton conductivity and stability of $Ba_2In_2O_5$ in hydrogen containing atmospheres. Journal of The Electrochemical Society, 158(1), B61.

[32] Su, X.T., Yan, Q.Z., Ma, X.H., Zhang, W.F. & Ge, C.C. (2006). Effect of co-dopant addition on the properties of yttrium and neodymium doped barium cerate electrolyte. Solid State Ionics, 177(11–12), 1041-1045.

[33] Dimos, D. & Mueller, C.H. (1998). Perovskite thin films for high-frequency capacitor applications. Annual Review of Materials Science, 28(1), 397–419.

[34] Subba Rao, T., Murthy, V.R.K. & Viswanathan, B. (1990). Review of perovskite ceramics—Microwave dielectric resonator materials. Ferroelectrics, 102(1), 155–160.

[35] Lunkenheimer, P., Bobnar, V., Pronin, A.V., Ritus, A.I., Volkov, A.A. & Loidl, A. (2002). Origin of apparent colossal dielectric constants. Physical Review B, 66(5), 052105.

[36] Sebald, J., Krohns, S., Lunkenheimer, P., Ebbinghaus, S.G., Riegg, S., Reller, A. & Loidl, A. (2010). Colossal dielectric constants: A common phenomenon in $CaCu_3Ti_4O_{12}$ related materials. Solid State Communications, 150(17–18), 857–860.

[37] Parida, K., Das, S., Mahapatra, P.K. & Choudhary, R.N.P. (2019). Relaxor behavior and impedance spectroscopic studies of chemically synthesized $SrCu_3Ti_4O_{12}$ ceramic. Materials Research Bulletin, 111, 7–16.

[38] Liang, P., Li, Y., Li, F., Chao, X. & Yang, Z. (2014). Effect of the synthesis route on the phase formation behavior and electric property of $Na_{0.5}Bi_{0.5}Cu_3Ti_4O_{12}$ ceramics. Materials Research Bulletin, 52, 42–49.

[39] Wang, Y., Jie, W., Yang, C., Wei, X. & Hao, J. (2019). Colossal permittivity materials as superior dielectrics for diverse applications. Advanced Functional Materials, 29(27), 1808118.

[40] Trainer, M. (2000). Ferroelectrics and the Curie-Weiss law. European Journal of Physics, 21(5), 459.

[41] Pascale, F., D'arco, P., Lacivita, V. & Dovesi, R. (2021). The superexchange mechanism in crystalline compounds. The case of KMF3 (M= Mn, Fe, Co, Ni) perovskites. Journal of Physics: Condensed Matter, 34(7), 074002.

[42] Ahmed, M.A. & El-Dek, S.I. (2006). Extraordinary role of Ca^{2+} ions on the magnetization of $LaFeO_3$ orthoferrite. Materials Science and Engineering: B, 128(1–3), 30–33.

[43] Kaur, P. & Singh, K. (2020). Review of perovskite-structure related cathode materials for solid oxide fuel cells. Ceramics International, 46(5), 5521–5535.

[44] Blanchard, P.E., Reynolds, E., Kennedy, B.J., Kimpton, J.A., Avdeev, M. & Belik, A.A. (2014). Anomalous thermal expansion in orthorhombic perovskite $SrIrO_3$: Interplay between spin-orbit coupling and the crystal lattice. Physical Review B, 89(21), 214106.

[45] Goodenough, J.B. & Zhou, J.S. (1998). Localized to itinerant electronic transitions in transition-metal oxides with the perovskite structure. Chemistry of Materials, 10(10), 2980–2993.

[46] Androulakis, J., Katsarakis, N. & Giapintzakis, J. (2001). Ferromagnetic and antiferromagnetic interactions in lanthanum cobalt oxide at low temperatures. Physical Review B, 64(17), 174401.

[47] Wang, F., Xie, Y., Chen, J., Fu, H. & Xing, X. (2013). First-principles study on negative thermal expansion of $PbTiO_3$. Applied Physics Letters, 103(22), 221901.

[48] Jiang, S., Hu, T., Gild, J., Zhou, N., Nie, J., Qin, M. & Luo, J. (2018). A new class of high-entropy perovskite oxides. Scripta Materialia, 142, 116–120.

[49] Zvereva, I.A., Popova, V.F., Vagapov, D.A., Toikka, A.M. & Gusarov, V.V. (2001). Kinetics of formation of Ruddlesden-Popper phases: I. Mechanism of $La_2SrAl_2O_7$ formation. Russian Journal of General Chemistry, 71, 1181–1185.

[50] Sarkar, A., Wang, Q., Schiele, A., Chellali, M.R., Bhattacharya, S.S., Wang, D. et al. (2019). High-entropy oxides: Fundamental aspects and electrochemical properties. Advanced Materials, 31(26), 1806236.

[51] Ma, J., Chen, K., Li, C., Zhang, X. & An, L. (2021). High-entropy stoichiometric perovskite oxides based on valence combinations. Ceramics International, 47(17), 24348–24352.

[52] Spiridigliozzi, L., Biesuz, M., Sglavo, V.M. & Dell'Agli, G. (2023). Design, synthesis and formation mechanism of a novel entropy-stabilized perovskite oxide derived from barium cerate/zirconate. Journal of the European Ceramic Society (in press).

[53] Dong, C., Wang, Q., Qiang, J.B., Wang, Y.M., Jiang, N., Han,G. & Xia, J.H. (2007). From clusters to phase diagrams: Composition rules of quasicrystals and bulk metallic glasses. Journal of Physics D: Applied Physics, 40(15), R273.

[54] Spiridigliozzi, L., Ferone, C., Cioffi, R. & Dell'Agli, G. (2023). Compositional design of single-phase rare-earth based high-entropy oxides (HEOs) by using the cluster-plus-glue atom model. Ceramics International, 49(5), 7662–7669.

[55] Wang, Y., Liu, J., Song, Y., Yu, J., Tian, Y., Robson, M.J. et al. (2023). High-entropy perovskites for energy conversion and storage: Design, synthesis, and potential applications. Small Methods, 2201138.

[56] Yan, J., Wang, D., Zhang, X., Li, J., Du, Q., Liu, X. et al. (2020). A high-entropy perovskite titanate lithium-ion battery anode. Journal of Materials Science, 55, 6942–6951.

Carbides and High-Entropy Carbides

Carbides can be categorized based on their structural formations into four primary groups: those containing individual carbon atoms, those with isolated pairs of carbon atoms, those with linear arrangements of carbon atoms, and those possessing a mesh-like network of carbon atoms [1].

In the family of alkali metals, the complexity of carbide structures formed increases from lithium to potassium, sodium, rubidium, and cesium. Lithium forms a simple carbide, Li_2C; sodium generates NaC_8, NaC_{16}, and NaC_{64}, in addition to Na_2C_2; and potassium forms a range of carbides including KC_8, KC_{16}, KC_{24}, KC_{36}, KC_{48}, and KC_{60} [2]. Rubidium and cesium follow similar trends except for the absence of a Cs_2C_2 compound. Such carbides display graphite-like layered arrangements with metal atoms situated between carbon layers.

The interspacing between the carbon layers varies, such as 0.540 nm for KC_8, KC_{24}, and KC_{36}; 0.565 nm for RbC_8, RbC_{24}, and RbC_{36}; 0.460 nm for NaC_{64}; and 0.594 nm for CsC_8.

In contrast, the alkali earth metals tend to form carbides with less complex anionic configurations, chiefly characterized by MeC_2-type structures (Me stands for a generic metal) that feature isolated carbon atom pairs. For example, the carbides CaC_2, SrC_2, and BaC_2 adopt a CaC_2-type distorted rock-salt structure with the C_2^{2-} units being parallel to one another (Fig. 1).

Lanthanides, as well as scandium and yttrium, produce carbides with compositions ranging from Me_3C, MeC, Me_2C_3 to MeC_2. Most rare earth metals, excluding lanthanum, cerium, praseodymium, and neodymium, form Me_3C carbides that adopt an Fe_3N-type cubic structure, akin to the NaCl-type face-centered cubic lattice but with fewer carbon atoms. MeC phases exist for yttrium, scandium, and cerium, which also align with the NaCl-type fcc lattice [3-5].

The sesquicarbides Me_2C_3 primarily adopt a Pu_2C_3-type [6] body-centered cubic structure, as has been reported for sesquicarbides of erbium, thulium, and lutetium [7-9]. The dicarbides crystallize in a CaC_2-type body-centered tetragonal structure containing two atoms per unit cell, and the lattice parameters vary according to the atomic number, showcasing the lanthanide contraction effect.

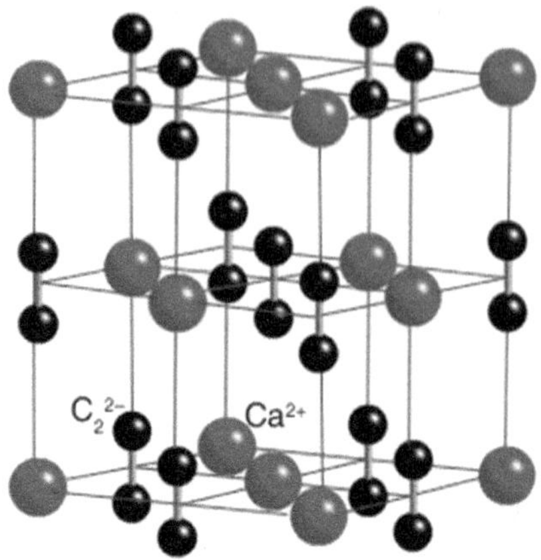

Fig. 1: Exemplary crystal structure of calcium carbide (CaC_2).

Nonmetallic carbide compounds garner attention for their high melting points, hardness, and resistance to abrasion and chemicals, along with their semiconducting capabilities. Noteworthy among these are the carbides of boron and silicon. B_4C and $B_{6.5}C$ have rhombohedral lattice structures [10, 11], while silicon carbide exists in cubic (β-SiC) and hexagonal/rhombohedral (α-SiC) forms [12, 13].

Alkali metal carbides display relative instability, undergoing decomposition even at temperatures around 800°C. However, as one progresses to carbides belonging to subgroups IVa, Va, and VIa, there is a marked elevation in melting points, reaching peak values in carbides like HfC, NbC, and TaC. It is noteworthy that the melting temperatures of these carbides generally exceed those of the corresponding metals, falling within a range of 0.55 to 0.98 times the melting point of the metal itself.

Carbides of titanium, zirconium, and hafnium evaporate without undergoing any compositional alterations [14, 15]. However, carbides of niobium and tantalum exhibit changes in their composition upon evaporation within the temperature range of 2770 to 3300°C, with carbon preferentially evaporating first, leading to non-equimolar compositions [16, 17].

A trend in the heats of formation for carbides exists: they generally decrease when transitioning from subgroup IVa metals to those in subgroups Va and VIa. This points towards a decrease in the metal-carbon bond energy, due to the influence of collective electrons, and a simultaneous increase in the strength of metal-metal and carbon-carbon covalent bonds. The weakening of the metal-carbon bond energy is ascribed to an increase in the statistical weight of stable d5-configurations when moving from subgroup IVa to Va and VIa carbides [2]. This reduces the probability of destabilizing the sp3-configurations of carbon and promotes a more pronounced separation between the metal and carbon sublattices within the carbides.

In terms of boiling points, carbides exhibit extremely high values, ranging from 2537°C for Be_2C to as high as 6000°C for WC [2].

The thermal expansion coefficients for transition metal carbides are comparable to those of the metals themselves and generally decrease with an increase in the atomic number of the constituent element.

Transition metal carbides, carbide phases with high carbon content in rare-earth metals, and carbides of nonmetals exhibit notable microhardness. This property tends to decrease when transitioning from carbides of metals in subgroup IVa to those in subgroups Va and VIa. Information regarding the microhardness of carbides in metals belonging to groups VII and VIII is currently limited. However, it is understood that these carbides are significantly less hard compared to those in subgroups IVa, Va, and VIa.

Specifically, the microhardness of such refractory materials, including carbides, can be attributed to the statistical likelihood of stable electron configurations in both the metal and carbon atoms that make up the compound. An increase in the statistical weight of stable sp3 configurations in carbon correlates with a rise in the resulting microhardness [18].

The unique attributes of carbides, particularly those of transition metals in Mendeleev's periodic table, pave the way for their extensive applicability in various technological domains. Characterized by elevated melting points, remarkable hardness, and resistance to both corrosion and wear, along with metallic-type electrical conduction, the carbides have evolved from their initial roles. While traditionally employed primarily as components in hard alloys, their application spectrum has expanded considerably in the past decade to include:

- Refractory Materials – Carbides can be utilized in high-temperature environments and industrial processes due to their outstanding heat resistance.
- Corrosion-Resistant Materials and Chemical Industry Applications – Carbides' resistance to chemical degradation makes them suitable for use in corrosive environments, as well as in chemical production processes.
- Nuclear Energy Production Materials – Given their high melting points and durability, carbides are ideal for applications involving nuclear reactions and high energy production.
- Aerospace Manufacturing Materials – Carbides are employed in the construction of aircraft and rocket components, leveraging their combination of light weight, high strength, and temperature resistance.
- Superhard and Wear-Resistant Materials – Carbides are used in situations demanding extreme hardness and durability, such as in cutting tools and abrasion-resistant coatings.

These versatile applications underscore the significant role that carbides play in advancing various fields of technology [2].

High-Entropy Carbides

In recent times, advancements in the realm of aerospace engineering have been intricately linked to strides made in the field of material sciences. When vehicles travel at hypersonic speeds, the thermal conditions on the leading edges and wings can escalate to as high as 2000°C [19]. In these demanding conditions, specialized materials capable of withstanding high temperatures are essential. Common choices include Silicon Carbide (SiC), Zirconium Carbide (ZrC), and Hafnium Carbide (HfC), often employed as strengthening agents or protective surface coatings to enhance the material's ability to operate under ultra-high-temperature (UHT) scenarios.

In 2004 [20], after the introduction of Yeh's concept regarding high-entropy alloys (HEAs), researchers began to explore the practicality of high-entropy carbides (HECs). Thus, in 2018, Yan et al. [21] and Castle et al. [22] reported the first instances of high-entropy carbides, including compounds like $(Hf_{0.25}Ta_{0.25}Zr_{0.25}Ti_{0.25})C$ and $(Hf_{0.2}Zr_{0.2}Ta_{0.2}Nb_{0.2}Ti_{0.2})C$, synthesized through Spark Plasma Sintering (SPS) using equimolar monocarbide powders.

Transition metal carbides like HfC, ZrC, TaC, NbC, TiC, and VC share a common rocksalt lattice structure. In line with Hume-Rothery theory, their atomic size misfit parameter is less than 10%, which theoretically suggests a propensity to form monophasic solid solutions. These carbides serve as core constituents in current high-entropy carbides (HECs). Additionally, though possessing hexagonal crystal structures, WC, MoC, W_2C, and Mo_2C can integrate into a face-centered cubic (FCC) solution when combined with the above-mentioned carbides, showing no preferred orientation.

MeC-type high-entropy carbides, such as $(Hf_{0.2}Zr_{0.2}Ta_{0.2}Nb_{0.2}Ti_{0.2})C$, manifest a rock-salt structural arrangement too, in which the carbon atoms are positioned at the (0,0,0) crystallographic sites, while the metallic elements are distributed at the (1/2,1/2,1/2) sites and are entirely disordered on a macroscopic scale [23]. Figure 2 shows the crystal structure of both monocarbides and high-entropy carbides.

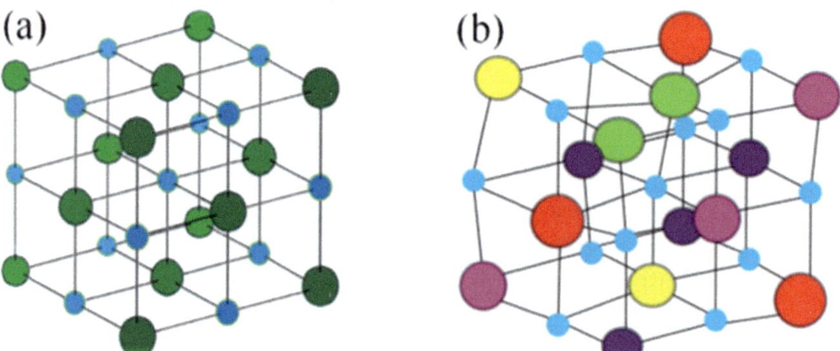

Fig. 2: Schematic crystal structure of monocarbides (a) and high-entropy carbides (b) [23].

Research focusing on HECs crystallography is scarce. Experimentally derived lattice constants for binary carbides and theoretical estimates for HECs generally concur, with minor deviations. One influencing factor is the presence of oxygen impurities; a lower oxygen content results in a greater lattice constant. Exploring the impact of lattice parameters during the formation of solid solutions is worthwhile. Castle et al. [22] discovered that the compound $(Hf_{0.25}Ta_{0.25}Zr_{0.25}Nb_{0.25})C$ more readily formed a monophase solid solution than $(Hf_{0.25}Ta_{0.25}Zr_{0.25}Ti_{0.25})C$, even though their lattice constants were almost identical.

In terms of their properties, however, HECs have been investigated by many researchers, indicating a reduction in thermal conductivity, enhancements in hardness levels, and improved resistance to oxidation as compared to traditional transition metal monocarbides like ZrC and TiC. These attributes earmark them as strong contenders for applications in extreme conditions, including but not limited to, fuel or cladding components in nuclear reactors or as part of the thermal protection systems (TPSs) in hypersonic vehicles designed for atmospheric re-entry.

Physiochemical Properties

To date, experimental data on the melting or decomposition thresholds for high-entropy carbides remain lacking. Liu et al. [24] took a computational approach, using first-principles calculations to estimate the melting points for a selection of quaternary metal carbides. Their projections indicate that 15 types of high-entropy carbides could potentially have melting points exceeding 3000 °C. Notably, the carbide composition $(Hf_{0.25}V_{0.25}Nb_{0.25}Ta_{0.25})C$ emerged as the most thermally robust, with a predicted melting point of 4400 °C. Preliminary thermal analyses have been performed on $(Hf_{0.2}Zr_{0.2}Ta_{0.2}Nb_{0.2}Ti_{0.2})C$ samples, subjecting them to annealing processes at four different temperature stages: 500 °C, 800 °C, 1140 °C, and 1700 °C. These sessions lasted for two hours each and were carried out in an argon atmosphere. Subsequent X-ray diffraction tests revealed no phase changes across this temperature range, implying robust thermal stability for this high-entropy carbide from ambient conditions up to 1700°C.

It is worth noting that high-entropy materials could possess elevated activation energies for diffusion. This implies that more prolonged annealing experiments may be necessary for a thorough examination of their stability or propensity for phase shifts or decomposition under extreme thermal conditions.

Mechanical Properties

The mechanical properties of high-entropy carbides (HECs) are paramount for their suitability in extreme environments. These characteristics are not merely a function of temperature but are also shaped by factors like residual porosity, grain size, and contamination with undesired oxides. Furthermore, mechanical property measurements such as hardness, strength, and fracture toughness can vary depending on the specific testing methods employed.

In general, the flexural strength of HECs could be measured using either a 3-point or 4-point bending configuration. Fracture toughness can be gauged using a variety of methodologies, including the Single-Edge V Notched Beam (SENVB), Chevron Notch Beam (CNB), and Indentation-Induced Crack Measurement (ICM). Among these, SENVB and CNB are considered more reliable compared to ICM, offering a broader and more robust data set for analysis [25].

Empirical data at room temperature indicate that high-entropy carbides manifest fracture toughness values ranging from 300 to 450 MPa and flexural strength values between $3 - 6$ MPa m$^{1/2}$ [26,27]. A marked improvement in both fracture toughness and flexural strength is noted when microstructural optimizations are made. For instance, a grain size reduction in $(Hf_{0.2}Zr_{0.2}Ta_{0.2}Nb_{0.2}Ti_{0.2})C$ from 16.5 to 0.6 μm leads to a 20% increase in fracture toughness and a 25% rise in flexural strength [28].

Overall, the mechanical integrity of high-entropy carbides is influenced by a confluence of factors, including microstructure, test methodologies, and temperature. To fully unlock their potential for high-temperature applications, these aspects still require careful consideration and optimization.

High-entropy carbides commonly exhibit high levels of hardness at room temperature, primarily measured using Vickers micro-indenters or Berkovich nano-indenters [29]. These materials often exhibit hardness values significantly above what would be estimated by the rule-of-mixtures [22]. The high hardness of HECs is attributed to strong metal-carbon covalent bonds and additional factors like mass inhomogeneity [30], solid solution hardening [22], atomic randomness at dislocation cores [31], and valence electron concentration (VEC) [32].

The following mechanisms for the HECs enhanced hardness have been proposed in literature:

- Mass Inhomogeneity: This generates an impedance mismatch along dislocation paths, increasing the material's resistance to plastic deformation [30].
- Solid Solution Hardening: Atomic size mismatch in the solid solution contributes to lattice distortion, thereby impeding the motion of dislocations essential for slip and plastic deformation [22].
- Dislocation Core Atomic Randomness: Random interactions among different elements at a dislocation core increase the resistance to dislocation slip, as corroborated by density functional theory (DFT) calculations of Peierls stress [31].
- Valence Electron Concentration (VEC): A regulating factor for the Fermi energy that affects bonding characteristics; it has been demonstrated that a VEC value equal to 8.4 leads to a maximum in HECs hardness [32].

Certain high-entropy carbides maintain their initial strength up to temperatures as high as 1800 °C. For example, $(Hf_{0.2}Zr_{0.2}Ta_{0.2}Nb_{0.2}Ti_{0.2})C$ ceramics exhibited a mean flexural strength of 421 MPa at room temperature, which remained relatively stable until 1800°C [33]. However, a subsequent decrease in strength was observed at temperatures beyond 1800°C, potentially due to mechanisms like creep deformation or a decrease in the elastic modulus.

Anyway, the decline in strength at temperatures exceeding 1800°C needs further scrutiny. This decrease is likely associated with creep deformation or the reduction of the elastic modulus. Studies like those by Demirskyi et al. [34,35] report similar trends, reinforcing the need for further investigation into the underlying mechanisms. While some high-entropy carbides show better strength retention at higher temperatures compared to monocarbides, the existing data is insufficient for a comprehensive comparison. The differences in flexural strength could be attributed to chemical bonding and compositional effects, which require further investigation.

In summary, the exceptional mechanical properties of high-entropy carbides and diborides are influenced by a complex interplay of factors including chemical bonding, atomic structure, and thermal stability. To fully comprehend their potential for high-temperature applications, a multi-faceted examination involving computational methods, empirical testing, and microstructural analysis is crucial.

Thermochemical Properties

In high-entropy carbides, the primary heat carriers are electrons and phonons [36]. These materials generally exhibit a reduced thermal conductivity compared to their monolithic counterparts.

The main factors influencing thermal conductivity in HECs are:

- Lattice Distortions: Due to the heterogeneous nature of the metal atoms in HECs, significant atomic-scale lattice distortions exist [37]. These distortions result in fluctuating mass and interatomic force constants that can lead to severe phonon scattering, thereby reducing thermal conductivity.
- Metal Element Count: Wang et al. [38] found that the thermal conductivity of transition metal carbides decreases as the number of metal elements increases.
- External Defects: In addition to lattice defects, thermal conductivity is also influenced by other defects like porosity, carbon stoichiometry [39], and irradiation-induced defects [40].

By altering the composition and controlling defects, it is possible to tailor the thermal properties of high-entropy carbides. For example, $(Hf_{0.25}Ta_{0.25}Zr_{0.25}Nb_{0.25})C$ exhibited thermal conductivity values that were generally lower than those of the constituent monocarbides [41, 42].

The contribution of electrons and phonons to thermal conductivity shows temperature dependency. An increase in thermal conductivity with temperature is mainly attributed to electron contribution, while the phonon contribution shows a weaker dependence on temperature.

The Coefficient of Thermal Expansion (CTE) is an essential factor in understanding the thermal stability of a material. For High-Entropy Carbides, available data is limited but reveals some intriguing trends: a study using a dilatometer showed that the thermal expansion coefficient of $(Hf_{0.2}Zr_{0.2}Ta_{0.2}Nb_{0.2}Ti_{0.2})C$ at room temperature was comparable to that of five monocarbides, namely ZrC, HfC, NbC, TiC, and TaC [43]. Moreover, Zhang et al. performed DFT calculations on $(Hf0.2Zr0.2Ta0.2Ti0.2Me0.2)C$, where Me = Cr, V, Nb, W, or Mo. The study concluded that the HEC with V has the smallest linear thermal expansion coefficient at high temperatures [44].

The observed anisotropy and temperature-dependent behavior underline the need for careful material design, especially when the material is expected to undergo substantial thermal fluctuations. Moreover, understanding the thermal expansion behavior can help in engineering materials that are less susceptible to thermal stress-induced failure.

Finally, given the limited data on thermal expansion coefficients for high-entropy carbides, there is a pressing need for more comprehensive studies that include temperature-dependent behavior, the influence of compositional complexity, and potential anisotropy in these materials.

Irradiation Resistance

Irradiation resistance is a critical factor when selecting materials for use in nuclear reactors and other environments subject to high levels of radiation.

Transition metal carbides, such as ZrC, have demonstrated their utility in a variety of nuclear applications, including gas-cooled fast reactors [45], high-temperature gas-cooled reactors [46], and space nuclear propulsion [47]. This preference stems from their high thermal conductivity, stability, and resistance to amorphization. HECs show comparable thermal conductivity to ZrC, making them candidates for fuel matrix and cladding applications.

The complexity in HECs arises from the mixing of multiple principal elements, which leads to atomic-level disorder and a heterogeneous potential energy landscape. This structural diversity affects the kinetics of radiation-induced defects, often suppressing their migration and clustering. Such behavior has been well-studied in high-entropy alloys [48-50], offering valuable insights into the potential radiation resistance of HECs.

Five key aspects influenced by chemical complexity in irradiation defect evolution can be identified:

  i.   The complexity can influence 3d electronic structures.
  ii. Atomic-level inhomogeneity and energy landscape are affected.
  iii. Non-adiabatic interactions between electrons and ions are influenced.
  iv. The pinning of point defects occurs.
  v.   Suppression of void swelling and helium bubble growth happens.

Early studies conducted on the irradiation resistance of HECs have revealed promising results; for example, $(Zr_{0.25}Ta_{0.25}Nb_{0.25}Ti_{0.25})C$ showed high phase stability even after exposure to 3 MeV Zr ions [51], as well as irradiation damage in $(W_{0.2}Ti_{0.2}V_{0.2}Nb_{0.2}Ta_{0.2})C$ showed dislocation loops at low doses and dislocation networks at higher doses [52].

There is an observed irradiation hardening effect in HECs, which leads to an 8%–10% increase in hardness compared to unirradiated samples. At the same time, radiation appears to have a greater impact on thermal conductivity at lower irradiation temperatures. Combining high irradiation resistance with other advantageous physical properties such as high strength and thermal conductivity, HECs offer a good materials option for advanced nuclear energy systems.

While the early findings are promising, more comprehensive studies are needed to ascertain the role of various irradiation-induced defects on the physical properties of HECs. Also, the impact of these defects on long-term stability, thermal properties, and mechanical properties under extreme conditions remains a key area for further investigation.

Definitely, high-entropy carbides present an intriguing class of materials with potential for use in environments requiring high irradiation resistance. Their complex structures and diverse compositions may offer unique mechanisms to suppress radiation-induced defects, thereby enhancing their applicability in demanding conditions such as those found in nuclear reactors.

Synthesis and Sintering Techniques

The synthesis of high-entropy carbides has primarily been accomplished through three distinct routes, depending on the initial raw materials used. These routes include: (1) utilizing commercially available metal carbide powders; (2) employing a combination of elemental metals and carbon powders; and (3) using metal oxide powders in conjunction with carbon powders.

The resulting HECs microstructure and elemental distribution are significantly impacted by the nature of the initial materials. Finer initial powders contribute to a more homogenous microstructure and elemental distribution, albeit with elevated levels of oxygen contamination. Throughout these synthesis pathways, a persistent issue is the presence of residual oxygen in the final sintered materials, a contamination that is challenging to completely remove during powder processing and high-temperature sintering.

Generally speaking, densification of HECs represents a formidable challenge, even under elevated temperature conditions. Here are presented some of the most promising sintering techniques investigated for possible HECs manufacturing:

- Hot Pressing (HP): Hot Pressing (HP) provides a means to achieve densification at comparatively lower temperatures without the need for additional sintering aids. HP also boasts the advantage of shorter sintering times, contributing to high-purity end products. Recent reports indicate that HP at 1850 °C yielded HECs with a relative density of 98.7% [53, 54]. However, HP is not without drawbacks. Grain growth remains a persistent issue, and the method tends to induce cracking. Hence, HP is typically considered suitable only for small, uncomplicated specimens.

- Spark Plasma Sintering (SPS): Spark Plasma Sintering (SPS) has been successfully employed to sinter HECs [55,56]. During the SPS process, a pulsed direct electric current flows through an electrically conductive piston-die setup, typically crafted from graphite, and the ceramic carbide or diboride powders. SPS shares similarities with HP, notably the use of external pressure. The unique attribute of SPS lies in the passage of a direct pulse current through the raw powder material. This mechanism prompts the individual particles to generate Joule heat and activate their surfaces. Subsequently, high-temperature plasma forms. The internal self-heating in SPS makes it more efficient than HP, cutting down the time required for sintering, as evidenced by available data. Nonetheless, SPS has limitations akin to HP, primarily its suitability for small and regular-shaped samples, which complicates post-treatment procedures.
- Selective Laser Sintering (SLS): The SLS process utilizes a high-power laser to selectively sinter a layer of powder material, effectively fusing the particles together to create a solid structure. The powders are heated through Joule heating, induced by the electric current in a vacuum setting. Zhang et al. [57] pioneered the use of SLS for ultra-fast reactive sintering of a non-equiatomic Zr-Nb-Hf-Ta-C composite, using a Yb fiber laser and a mixture of monocarbides.
- Pressureless Sintering: Pressureless sintering has been suggested as a potential method that can achieve densification under ambient atmospheric conditions. Given the inherently slow self-diffusion rate of ceramics, this method necessitates the use of specific sintering aids and must be performed under high-temperature conditions within an inert gas atmosphere, typically argon. To date, there are no published reports on successful pressureless sintering of HECs. However, previous works on doped carbides have been undertaken by certain research groups, including the successful fabrication of $Ta_{0.8}Hf_{0.2}C$ [58] through pressureless sintering, with promising properties, and further reports on pressureless sintering of HECs are underway.

Definitely, to enhance the mechanical and thermal properties of high-entropy carbides, it's vital to reduce porosity, refine grain size, and eliminate contaminations. Controlling the purity and particle size of the starting powders is crucial for achieving these objectives. Smaller particle sizes facilitate lower sintering temperatures, thereby reducing grain growth during the process. Additionally, meticulous control over impurities, especially oxygen, is vital.

Machine Learning Models to Predict the Formation and the Properties of High-Entropy Carbides

The burgeoning interest in High-Entropy Carbides is justified by their intriguing phase formation abilities and promising applications. While the science of HECs is still in nascent stages, multiple approaches, empirical and computational, are being employed to understand the underlying mechanisms dictating their phase stabilities.

Traditionally, phase stability in HECCs has been understood through the lens of Gibbs free energy and mixing enthalpy. However, this understanding appears to be limited as no discernable threshold can be isolated for $\Delta H_{mix}/\Delta S_{mix}$ or ΔH_{mix} between single-phase and multi-phase samples. Even well-established empirical rules like the Ω-δ criterion [59], effective in the realm of high-entropy alloys [60], exhibit limitations when extrapolated to HEC systems. For instance, samples with volume misfit (δ) exceeding the proposed criteria were still observed to form single phases.

The advent of machine learning (ML) models in predicting the phase stability of High-Entropy Carbides marks a transformative milestone in materials science research. Zhang's work [61] demonstrates the efficacy of employing well-trained Support Vector Machine (SVM) and Artificial Neural Network (ANN) models for this endeavor. These models are trained using density functional theory (DFT) calculations as well as chemical attributes of HECC candidates and their constituting Transition Metal Carbides (TMCs).

It is well-understood that TMCs from Group IV and V readily form single-phase HECs with enhanced mechanical properties [2]. However, the inclusion of metals from Group VI (e.g. Chromium, Molybdenum, and Tungsten) introduces complexity in phase stability, largely owing to their higher valence filling. Despite this complication, the high valence filling in Group VI metals is anticipated to improve the overall performance of HECs.

Zhang's ML models serve as efficient tools for the fast screening of elemental combinations conducive to forming single-phase HECCs. This addresses the critical need for high-throughput methods for material discovery and design. Importantly, these models are not merely theoretical constructs; they have undergone rigorous validation through experimental work. This empirical validation provides a compelling testament to the model's utility, as several new HECs were successfully discovered and characterized as per the model's predictions.

Zhang's study goes one step further by applying the refined ML model to assess the single-phase formation probability of nonequiatomic HECs. The high level of agreement between the model's predictions and the experimental results substantiates the applicability of ML in this complex domain, possibly accelerating the discovery process of new functional HECs by circumventing the challenges of traditional trial-and-error methods and computational limitations.

Definitely, the incorporation of machine learning techniques into the field of high-entropy carbides presents a monumental shift towards a data-driven paradigm. Anyway, the ability to effectively design and tune the properties of new HECs through a data-driven approach is still far from being mature.

References

[1] Hausner, H.H. (2013). Coatings of High-Temperature Materials. Springer.

[2] Kosolapova, T.Y. (2012). Carbides: Properties, Production, and Applications. Springer Science & Business Media.

[3] Maibam, J., Sharma, B.I., Bhattacharjee, R., Thapa, R.K. & Singh, R.B. (2011). Electronic structure and elastic properties of scandium carbide and yttrium carbide: A first principles study. Physica B: Condensed Matter, 406(21), 4041–4045.

[4] Abu-Jafar, M.S., Leonhardi, V., Jaradat, R., Mousa, A.A., Al-Qaisi, S., Mahmoud, N.T. & Bouhemadou, A. (2021). Structural, electronic, mechanical, and dynamical properties of scandium carbide. Results in Physics, 21, 103804.

[5] Wang, W., Xue, S., Li, J., Wang, F., Kang, Y. & Lei, Z. (2017). Cerium carbide embedded in nitrogen-doped carbon as a highly active electrocatalyst for oxygen reduction reaction. Journal of Power Sources, 359, 487–493.

[6] Matthews, R.B. & Herbst, R.J. (1983). Uranium-plutonium carbide fuel for fast breeder reactors. Nuclear Technology, 63(1), 9–22.

[7] Dilmi, S., Saib, S. & Bouarissa, N. (2018). Electron-phonon interaction in the binary superconductor lutetium carbide LuC2 via first-principles calculations. Physica C: Superconductivity and its Applications, 549, 131–138.

[8] Gschneidner, K.A. & Calderwood, F.W. (1986). The C-Tm (Carbon-Thulium) system. Bulletin of Alloy Phase Diagrams, 7(6), 563–564.

[9] Ahn, H.J., Moon, J., Koh, S., Seo, Y., Kim, C.K., Rho, I.C. & Cho, B.J. (2016). Very low-work-function ALD-Erbium Carbide (ErC 2) metal electrode on high-$ K $ dielectrics. IEEE Transactions on Electron Devices, 63(7), 2858–2863.

[10] Manohar, G., Pandey, K.M. & Maity, S.R. (2021). Characterization of Boron Carbide (B4C) particle reinforced aluminium metal matrix composites fabricated by powder metallurgy techniques–A review. Materials Today: Proceedings, 45, 6882–6888.

[11] He, Q., Tian, S., Xie, J., Xiang, C., Wang, H., Wang, W. & Fu, Z. (2020). Microstructure and anisotropic mechanical properties of B6.5C-TiB2-SiC-BN composites fabricated by reactive hot pressing. Journal of the European Ceramic Society, 40(8), 2862–2869.

[12] Zhou, X.T., Lai, H.L., Peng, H.Y., Au, F.C., Liao, L.S., Wang, N. & Lee, S.T. (2000). Thin β-SiC nanorods and their field emission properties. Chemical Physics Letters, 318(1–3), 58–62.

[13] Starke, U. (1997). Atomic structure of hexagonal SiC surfaces. Physica Status Solidi (b), 202(1), 475–499.

[14] Torshina, V.V., Smolina, G.N., Dobychin, S.L., Avarbe, R.G. & Vil'k, Y.N. (1974). Mass-spectrometric study of the evaporation of zirconium carbide at high temperatures. Refractory Carbides, 335–339.

[15] Fesenko, V.V. & Bolgar, A.S. (1969). Study of the evaporation rates of the carbides of titanium, zirconium, hafnium, niobium, and tantalum at high temperatures. Institute for Problems in Materials Science, Kiev, Ukraine.

[16] Fries, R.J. (1962). Vaporization behavior of niobium carbide. The Journal of Chemical Physics, 37(2), 320–327.

[17] Eckstein, B.H. & Forman, R. (1962). Preparation and some properties of tantalum carbide. Journal of Applied Physics, 33(1), 82–87.

[18] Samsonov, G.V. (1974). Some problems in the theory of the properties of carbides. Refractory Carbides, 1–11.

[19] Silvestroni, L., Kleebe, H.J., Fahrenholtz, W.G. & Watts, J. (2017). Super-strong materials for temperatures exceeding 2000°C. Scientific Reports, 7(1), 40730.

[20] Yeh, J.W., Chen, S.K., Lin, S.J., Gan, J.Y., Chin, T.S., Shun, T.T. & Chang, S.Y. (2004). Nanostructured high-entropy alloys with multiple principal elements: Novel alloy design concepts and outcomes. Advanced Engineering Materials, 6(5), 299–303.

[21] Yan, X., Constantin, L., Lu, Y., Silvain, J.F., Nastasi, M. & Cui, B. (2018). $(Hf_{0.2}Zr_{0.2}Ta_{0.2}Nb_{0.2}Ti_{0.2})C$ high-entropy ceramics with low thermal conductivity. Journal of the American Ceramic Society, 101(10), 4486–4491.

[22] Castle, E., Csanádi, T., Grasso, S., Dusza, J. & Reece, M. (2018). Processing and properties of high-entropy ultra-high temperature carbides. Scientific Reports, 8(1), 8609.

[23] Yu, D., Yin, J., Zhang, B., Liu, X. & Huang, Z. (2020). Recent development of high-entropy transitional carbides: Areview. Journal of the Ceramic Society of Japan, 128(7), 329–335.

[24] Liu, B., Zhao, J., Liu, Y., Xi, J., Li, Q., Xiang, H. & Zhou, Y. (2021). Application of high-throughput first-principles calculations in ceramic innovation. Journal of Materials Science & Technology, 88, 143–157.

[25] Quinn, G.D. & Bradt, R.C. (2007). On the Vickers indentation fracture toughness test. Journal of the American Ceramic Society, 90(3), 673–680.

[26] Nisar, A., Zhang, C., Boesl, B. & Agarwal, A. (2020). A perspective on challenges and opportunities in developing high entropy-ultra high temperature ceramics. Ceramics International, 46(16), 25845–25853.

[27] Feng, L., Fahrenholtz, W.G. & Brenner, D.W. (2021). High-entropy ultra-high-temperature borides and carbides: Anew class of materials for extreme environments. Annual Review of Materials Research, 51, 165–185.

[28] Wang, F., Zhang, X., Yan, X., Lu, Y., Nastasi, M., Chen, Y. & Cui, B. (2020). The effect of submicron grain size on thermal stability and mechanical properties of high-entropy carbide ceramics. Journal of the American Ceramic Society, 103(8), 4463–4472.

[29] Wang, F., Monteverde, F. & Cui, B. (2023). Will high-entropy carbides and borides be enabling materials for extreme environments? International Journal of Extreme Manufacturing, 5(2), 022002.

[30] Sarker, P., Harrington, T., Toher, C., Oses, C., Samiee, M., Maria, J.P. & Curtarolo, S. (2018). High-entropy high-hardness metal carbides discovered by entropy descriptors. Nature Communications, 9(1), 4980.

[31] Wang, Y., Csanádi, T., Zhang, H., Dusza, J., Reece, M.J. & Zhang, R.Z. (2020). Enhanced hardness in high-entropy carbides through atomic randomness. Advanced Theory and Simulations, 3(9), 2000111.

[32] Hossain, M.D., Lowum, S., Borman, T. & Maria, J.P. (2021). Fermi level engineering and mechanical properties of high entropy carbides. arXiv preprint arXiv:2101.04885.

[33] Feng, L., Chen, W.T., Fahrenholtz, W.G. & Hilmas, G.E. (2021). Strength of single-phase high-entropy carbide ceramics up to 2300°C. Journal of the American Ceramic Society, 104(1), 419–427.

[34] Demirskyi, D., Borodianska, H., Suzuki, T.S., Sakka, Y., Yoshimi, K. & Vasylkiv, O. (2019). High-temperature flexural strength performance of ternary high-entropy carbide consolidated via spark plasma sintering of TaC, ZrC and NbC. Scripta Materialia, 164, 12–16.

[35] Demirskyi, D., Suzuki, T.S., Yoshimi, K. & Vasylkiv, O. (2020). Synthesis and high-temperature properties of medium-entropy (Ti, Ta, Zr, Nb)C using the spark plasma consolidation of carbide powders. Open Ceramics, 2, 100015.

[36] Zhang, P.X., Ye, L., Chen, F.H., Han, W.J., Wu, Y.H. & Zhao, T. (2022). Stability, mechanical, and thermodynamic behaviors of (TiZrHfTaM)C (M = Nb, Mo, W, V, Cr) high-entropy carbide ceramics. Journal of Alloys and Compounds, 903, 163868.

[37] Monteverde, F., Gaboardi, M., Saraga, F., Feng, L., Fahrenholtz, W. & Hilmas, G. (2022). Anisotropic thermal expansion in high-entropy multicomponent AlB2-type diboride solid solutions. International Journal of Extreme Manufacturing, 5(1), 015505.

[38] Wang, Y. (2022). Processing and properties of high entropy carbides. Advances in Applied Ceramics, 121(2), 57–78.

[39] Rost, C.M., Borman, T., Hossain, M.D., Lim, M., Quiambao-Tomko, K.F., Tomko, J.A. & Hopkins, P.E. (2020). Electron and phonon thermal conductivity in high entropy carbides with variable carbon content. Acta Materialia, 196, 231–239.

[40] Dennett, C.A., Hua, Z., Lang, E., Wang, F. & Cui, B. (2022). Thermal conductivity reduction in $(Zr_{0.25}Ta_{0.25}Nb_{0.25}Ti_{0.25})C$ high entropy carbide from extrinsic lattice defects. Materials Research Letters, 10(9), 611–617.

[41] Schwind, E.C., Reece, M.J., Castle, E., Fahrenholtz, W.G. & Hilmas, G.E. (2022). Thermal and electrical properties of a high entropy carbide (Ta, Hf, Nb, Zr) at elevated temperatures. Journal of the American Ceramic Society, 105(6), 4426–4434.

[42] Zhou, Y., Fahrenholtz, W.G., Graham, J. & Hilmas, G.E. (2021). From thermal conductive to thermal insulating: Effect of carbon vacancy content on lattice thermal conductivity of ZrCx. Journal of Materials Science & Technology, 82, 105–113.

[43] Yan, X., Constantin, L., Lu, Y., Silvain, J.F., Nastasi, M. & Cui, B. (2018). $(Hf_{0.2}Zr_{0.2}Ta_{0.2}Nb_{0.2}Ti_{0.2})C$ high-entropy ceramics with low thermal conductivity. Journal of the American Ceramic Society, 101(10), 4486–4491.

[44] Zhang, P.X., Ye, L., Chen, F.H., Han, W.J., Wu, Y.H. & Zhao, T. (2022). Stability, mechanical, and thermodynamic behaviors of (TiZrHfTaM)C (M = Nb, Mo, W, V, Cr) high-entropy carbide ceramics. Journal of Alloys and Compounds, 903, 163868.

[45] Ueta, S., Aihara, J., Sawa, K., Yasuda, A., Honda, M. & Furihata, N. (2011). Development of high temperature gas-cooled reactor (HTGR) fuel in Japan. Progress in Nuclear Energy, 53(7), 788–793.

[46] Ueta, S., Aihara, J., Yasuda, A., Ishibashi, H., Takayama, T. & Sawa, K. (2008). Fabrication of uniform ZrC coating layer for the coated fuel particle of the very high temperature reactor. Journal of Nuclear Materials, 376(2), 146–151.

[47] Hamilton, S., Jerred, N.D., Scott, R., Bachhav, M., Yao, T. & Miller, V.M. (2023). Diffusion study of uranium mononitride/zirconium carbide composite for space nuclear propulsion. Journal of Nuclear Materials, 583, 154535.

[48] Xia, S.Q., Zhen, W.A.N.G., Yang, T.F. & Zhang, Y. (2015). Irradiation behavior in high entropy alloys. Journal of Iron and Steel Research, International, 22(10), 879–884.

[49] Xu, Q., Guan, H.Q., Zhong, Z.H., Huang, S.S. & Zhao, J.J. (2021). Irradiation resistance mechanism of the CoCrFeMnNi equiatomic high-entropy alloy. Scientific Reports, 11(1), 608.

[50] Cheng, Z., Sun, J., Gao, X., Wang, Y., Cui, J., Wang, T. & Chang, H. (2023). Irradiation effects in high-entropy alloys and their applications. Journal of Alloys and Compounds, 930, 166768.

[51] Wang, F., Yan, X., Wang, T., Wu, Y., Shao, L., Nastasi, M., ... & Cui, B. (2020). Irradiation damage in $(Zr_{0.25}Ta_{0.25}Nb_{0.25}Ti_{0.25})C$ high-entropy carbide ceramics. Acta Materialia, 195, 739–749.

[52] Zhu, Y., Chai, J., Wang, Z., Shen, T., Niu, L., Li, S., ... & Cui, M. (2022). Microstructural damage evolution of $(WTiVNbTa)C5$ high-entropy carbide ceramics induced by self-ions irradiation. Journal of the European Ceramic Society, 42(6), 2567–2576.

[53] Monteverde, F. & Saraga, F. (2020). Entropy stabilized single-phase (Hf, Nb, Ta, Ti, Zr) B2 solid solution powders obtained via carbo/boro-thermal reduction. Journal of Alloys and Compounds, 824, 153930.

[54] Monteverde, F., Saraga, F. & Gaboardi, M. (2020). Compositional disorder and sintering of entropy stabilized (Hf, Nb, Ta, Ti, Zr)B2 solid solution powders. Journal of the European Ceramic Society, 40(12), 3807–3814.

[55] Ye, B., Wen, T., Nguyen, M.C., Hao, L., Wang, C.Z. & Chu, Y. (2019). First-principles study, fabrication and characterization of $(Zr_{0.25}Nb_{0.25}Ti_{0.25}V_{0.25})C$ high-entropy ceramics. Acta Materialia, 170, 15–23.

[56] Harrington, T.J., Gild, J., Sarker, P., Toher, C., Rost, C.M., Dippo, O.F. & Vecchio, K.S. (2019). Phase stability and mechanical properties of novel high entropy transition metal carbides. Acta Materialia, 166, 271–280.

[57] Zhang, X., Li, N., Chen, X., Stroup, M., Lu, Y. & Cui, B. (2023). Direct selective laser sintering of high-entropy carbide ceramics. Journal of Materials Research, 38(1), 187–197.

[58] Ghaffari, S.A., Faghihi-Sani, M.A., Golestani-Fard, F. & Ebrahimi, S. (2013). Pressureless sintering of $Ta_{0.8}Hf_{0.2}C$ UHTC in the presence of $MoSi_2$. Ceramics International, 39(2), 1985–1989.

[59] Liu, S.Y. et al. Phase stability, mechanical properties and melting points of high entropy quaternary metal carbides from first-principles. Journal of the European Ceramic Society, 41, 6267–6274.

[60] Yang, X. & Zhang, Y. (2012). Prediction of high-entropy stabilized solid-solution in multicomponent alloys. Materials Chemistry and Physics, 132, 233–238.

[61] Zhang, J., Xu, B., Xiong, Y., Ma, S., Wang, Z., Wu, Z. & Zhao, S. (2022). Design high-entropy carbide ceramics from machine learning. NPJ Computational Materials, 8(1), 5.

Borides and High-Entropy Borides

Borides constitute a fascinating category of materials with a diverse array of potential uses, due to their unique combination of mechanical, optical, electronic, and heat-resistant properties. These varying physical characteristics can be attributed to each boride's specific crystal configuration, which is fundamentally shaped by how the boron atoms are organized.

The metal-to-boron ratio in these compounds typically spans from 4:1 (Me_4B) to 1:12 (MeB_{12}), although there are some outliers like MeB_{50} and MeB_{66}.

These compounds can be formed with alkali metals (e.g. Li, Na, K), alkaline earth metals, as well as transition metals from groups III through VIII, in addition to lanthanides and actinides like Thorium, Uranium, and Plutonium [1].

Kiessling et al. [2] proposed a classification system for boron atom arrangements within borides, which can be categorized as follows: individual boron atoms, boron atom pairs, linear arrays of boron atoms, dual linear arrays, lattice structures made of boron atoms, and foundational frameworks consisting of boron atoms (see Figure 1).

In certain borides with a boron atom percentage exceeding 50%, the boron atoms create a wavy net-like structure, exemplified by types like CrB_4 and MnB_4, or form B6 octahedral structures, as seen in CaB_6. In cases where the boron content surpasses 85 atom%, the foundational framework is constructed of cuboctahedral (UB_{12}, ScB_{12}) or icosahedral configurations (AlB_{12}, YB_{66}).

The diversity in structure and bonding within metal borides naturally leads to a broad spectrum of different properties. Interestingly, despite this wide range of potential uses, their commercial applications remain rather limited. In the following sections, a brief overview of some key properties and potential applications of metal borides is given.

Catalysts

Since the latter half of the 20th century, powdered forms of metal borides have found utility as catalysts in various organic transformations, including hydrogenation and reduction reactions [3-5]. Additionally, due to their capacity

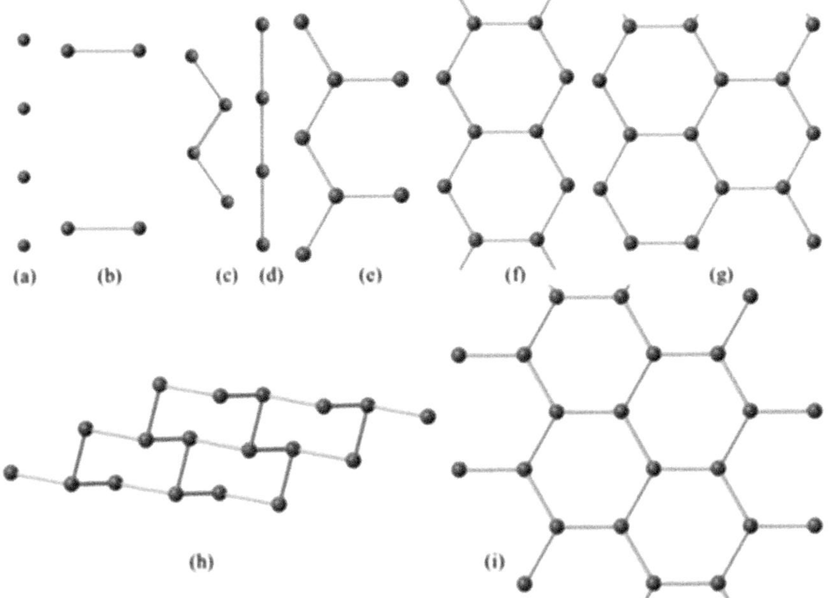

Fig. 1: Different structures adopted by boron atoms in metal borides: (a) isolated boron atoms; (b) pairs of boron atoms; (c) zig-zag chains of boron atoms; (d) straight (ordinary) chains of boron atoms; (e) extended chains of boron atoms; (f) double chains of boron atoms; (g) triple chains of boron atoms; (h) corrugated networks of boron atroms; (i) flat networks of boron atoms [1].

to catalytically decompose borohydrides in aqueous solutions, they have seen applications in fuel cells as hydrogen producers. Although traditionally synthesized through the precipitation of metallic salts using sodium borohydride, the actual mechanisms behind their catalytic behavior were not fully elucidated until recently. These borides are typically prepared in solution at low temperatures [6], resulting in an amorphous structural state.

Hard and Superhard Materials

The tendency of boron atoms to link together frequently gives rise to expansive three-dimensional networks characterized by covalent bonds. Such an arrangement contributes to a high shear modulus, as dislocations find it challenging to pass through these robust, directionally-oriented bonds. The incorporation of ultra-incompressible metals into the structure of borides yields compounds that rival the hardness and incompressibility of diamond, inspiring the creation of superhard metal borides [7-9].

Currently, the primary application of these hard borides is in machining tools. However, the exceptional wear resistance they offer opens up possibilities for additional uses, such as in the fabrication of prosthetic devices [10].

High-Temperature Materials

Given that boron has an outstanding melting point among single elements [11], it is logical that a substantial number of metal borides are heat-resistant or refractory in nature.

Noteworthy among these is zirconium diboride ($T_{melting}$ = 3230 °C), which features a two-dimensional honeycomb boron structure crystallizing in a configuration akin to that of AlB_2. The material's high melting point is complemented by an impressive elastic modulus, attributed to its flexible boron layers, making it a robust material capable of withstanding extreme thermal conditions. Furthermore, zirconium diboride demonstrates exceptional resistance to oxidation. It can endure thermal cycling conditions up to 2700 °C in an atmospheric setting. Oxidation byproducts for this material are ZrO_2 and B_2O_3; it is noteworthy that substantial oxidation only initiates above 1000 °C. This is largely due to the sublimation of B_2O_3, which reveals new surfaces that are more susceptible to degradation [1].

The convergence of these properties makes zirconium diboride an ideal candidate for applications demanding resilience to high-temperature environments [12, 13].

Superconductors

Given the rich array of bonding configurations observed in metal borides, it is not unexpected to find a number of promising superconductor candidates within this class [14-16]. Magnesium diboride is particularly noteworthy, not only because it serves as an effective precursor in generating transition metal borides via solid-state metathesis reactions [1], but also due to its remarkable property of undergoing a resistivity transition at 39 K. This temperature surpasses those observed in all type-I superconductors and even eclipses $La_{2-x}Ba_xCuO_4$, which holds the distinction of being the first discovered type-II superconductor [17].

Kondo Insulators

Although many rare earth borides exhibit superconducting behavior at low temperatures, there exists a distinct subset that becomes insulating, known as Kondo insulators [18]. This unique state arises in rare earth borides where the rare earth elements display mixed valency (for example, Yb can exist in either Yb^{+2} or Yb^{+3} oxidation states). The consequence of this mixed valency is the emergence of a band gap at reduced temperatures. For instance, in YbB_{12}, the valence band consists of 36 electrons contributed by the boron lattice, along with two electrons from the metallic component. An additional electron originating from a trivalent rare earth fills the conduction band exclusively.

It is therefore understandable that certain borides, due to their strong coupling and mixed valency in f-orbitals, are identified as ideal Kondo insulators [19, 20].

Samarium hexaboride, verified through resistivity and magnetic measurements, was one of the inaugural and most studied Kondo insulators [21].

Optical Properties of Borides

While metal borides typically lack any striking optical features in the visible spectrum—except for different coloration—they do exhibit significant interactions with photons at other wavelengths. Such interactions have paved the way for specialized applications [1, 22].

Thin boride films, for instance, have been employed as ultraviolet light filters, a functionality that could find use in advanced deep-space telescopes [23].

As another example, lanthanum hexaboride has a designated role in crystallography, serving as a reliable X-ray standard known as SRM 660a [24]. This underlines the versatility of borides in both common and highly specialized applications, expanding the scope of their utility beyond what might be traditionally expected.

High-Entropy Borides

In 2016, Gild et al. [25] reported the first-ever fabrication of single-phase high-entropy diborides (HEBs) with an AlB_2-prototyped hexagonal structure. They synthesized several HEB compositions such as $(Hf_{0.2}Zr_{0.2}Ta_{0.2}Nb_{0.2}Ti_{0.2})B_2$ and $(Hf_{0.2}Zr_{0.2}Ta_{0.2}Mo_{0.2}Ti_{0.2})B_2$ using high-energy ball milling of individual commercial boride powders, subsequently followed by spark plasma sintering (SPS) at 2000 °C under a uniaxial pressure of 30 MPa. However, one challenge encountered in this study was the presence of oxide impurities in the sintered ceramics. These impurities originated from surface oxides present on the commercial powders used. The oxide levels were found to increase during the ball milling process, and their presence acted as a barrier to achieving full densification, leading to ceramics with lower relative densities, approximately around 92%.

Following the pioneering research of Gild et al., subsequent studies have explored alternative routes for synthesizing these intriguing materials. Some employed boro/carbothermal reduction of metal oxides [26, 27], while others opted for elemental precursors [28, 29]. These researches point to the critical role that high-purity HEB powders with fine particle sizes play in the successful fabrication of dense, high-quality HEBs at reduced sintering temperatures. The work of Feng's et al. [30] was directed towards this particular aspect, aiming to synthesize high-purity HEB powders that could be employed to manufacture densely packed, high-quality $(Hf_{0.2}Zr_{0.2}Ti_{0.2}Ta_{0.2}Nb_{0.2})B_2$ ceramics with fine grain sizes. Feng's findings imply that optimizing the purity and particle size of initial boride powders could serve as a vital strategy for achieving high-quality high-entropy borides, thereby enhancing their potential for various practical applications.

Anyway, most of the first published studies on high-entropy borides were confined to diborides, with only a limited number of investigations exploring high-entropy borides with other metal-to-boron stoichiometric ratios such as monoborides (MeB) or hexaborides (MeB_6).

A noteworthy contribution came from Qin et al. [31], who successfully fabricated single-phase high-entropy monoborides, including compositions such as $(V_{0.2}Cr_{0.2}Nb_{0.2}Ta_{0.2}W_{0.2})B$ and $(V_{0.2}Cr_{0.2}Nb_{0.2}Mo_{0.2}Ta_{0.2})B$, using a novel method of reactive spark plasma sintering (SPS) that yielded bulk specimens with impressively high relative densities ranging from ~98.3% to 99.5%. Moreover, the measured Vickers hardness of Qin's high-entropy monoborides ranges from ~22-26 GPa at an indentation load of 9.8 N and escalate to ~32-37 GPa at a reduced indentation load of 0.98 N. Particularly, $(V_{0.2}Cr_{0.2}Nb_{0.2}Ta_{0.2}W_{0.2})B$ stands out for its exceptional hardness, exceeding even the values for ternary $(Ta_{0.5}W_{0.5})B$, which is already considered as superhard [32].

Other than monoborides, the *Immm* orthorhombic structure (space group No. 71) of Ta_3B_4-prototyped Me_3B_4 compounds (where Me represents a transition metal like Ti, V, Cr, Mn, Nb, and Ta) is particularly interesting. It features layers of metal cations and boron chains, separated by monolayers of metal cations. These Me_3B_4 phases, known to possess both high melting points and hardness, are predicted to exhibit excellent thermodynamic stability and elastic properties.

Qin et al. [33] extended the utility of the direct boron-metal reactive SPS method (already applied to monoborides) to synthesize high-entropy Me_3B_4 borides, specifically $(V_{0.2}Cr_{0.2}Nb_{0.2}Mo_{0.2}Ta_{0.2})_3B_4$ and $(V_{0.2}Cr_{0.2}Nb_{0.2}Ta_{0.2}W_{0.2})_3B_4$, with only 4%–5% of secondary phases. Both samples exhibited high relative densities of approximately 98.7%, and Vickers microhardness values ranging from 18.6 to 19.9 GPa at a 9.8 N indentation load. While these values are lower compared to those of high-entropy monoborides, they are comparable to high-entropy diborides [27, 29]. This places the high-entropy Me_3B_4 borides within an interesting hardness range that might be desirable for specific applications where a balance between hardness and other mechanical or thermal properties is required.

Finally, Qin et al. [34] also synthesized several bulk MeB6-typed cubic rare earth high-entropy hexaborides (where Me stands for a combination of five different rare earth metal among Y, La, Pr, Nd, Sm, Gd, Tb, Dy, or Yb). From a mechanical standpoint, these high-entropy hexaborides show impressive Vickers microhardness values ranging between 16 and 18 GPa at a standard indentation load of 9.8 N. Additionally, the nanoindentation hardness and Young's moduli are measured to range among 19-22 GPa and 190-250 GPa, respectively. Moreover, one of the most intriguing aspects of Qin's hexaborides is their elevated work function values, determined to be between 3.7 and 4.0 eV through ultraviolet photoelectron spectroscopy, being significantly higher than that of LaB_6, which could open new avenues in electronic applications requiring high work functions, such as electron emitters in electron microscopy or thermionic energy converters.

Definitely, the recent findings on high-entropy borides are not only substantial contributions to the burgeoning field of high-entropy materials but also represent a significant step towards optimizing the balance between hardness, density, and purity in borides. However, further work is needed to delve into a more detailed understanding of the mechanical and thermal properties of such high-entropy borides, especially to explore any possible practical applications exploiting their unique property combinations.

References

[1] Akopov, G., Yeung, M.T. & Kaner, R.B. (2017). Rediscovering the crystal chemistry of borides. Advanced Materials, 29(21), 1604506.

[2] Kiessling, R., Samuelson, O., Lindstedt, G. & Kinell, P.O. (1950). The borides of manganese. Acta Chemica Scandinavica, 4, 146–159.

[3] Chen, Y.Z. & Chen, Y.C. (1994). Hydrogenation of para-chloronitrobenzene over nickel borides. Applied Catalysis A: General, 115(1), 45–57.

[4] Deshpande, V.M., Ramnarayan, K. & Narasimhan, C.S. (1990). Studies on ruthenium-tin boride catalysts II. Hydrogenation of fatty acid esters to fatty alcohols. Journal of Catalysis, 121(1), 174–182.

[5] Collins, D.J., Smith, A.D. & Davis, B.H. (1982). Hydrogenation of nitrobenzene over a nickel boride catalyst. Industrial & Engineering Chemistry Product Research and Development, 21(2), 279–281.

[6] Chen, H. & Zou, X. (2020). Intermetallic borides: Structures, synthesis and applications in electrocatalysis. Inorganic Chemistry Frontiers, 7(11), 2248–2264.

[7] Levine, J.B., Tolbert, S.H. & Kaner, R.B. (2009). Advancements in the search for superhard ultra-incompressible metal borides. Advanced Functional Materials, 19(22), 3519–3533.

[8] Mohammadi, R., Turner, C.L., Xie, M., Yeung, M.T., Lech, A.T., Tolbert, S.H. & Kaner, R.B. (2016). Enhancing the hardness of superhard transition-metal borides: Molybdenum-doped tungsten tetraboride. Chemistry of Materials, 28(2), 632–637.

[9] Akopov, G., Pangilinan, L.E., Mohammadi, R. & Kaner, R.B. (2018). Perspective: Superhard metal borides: A look forward. APL Materials, 6(7).

[10] Kanyanta, V. (2016). Hard, superhard and ultrahard materials: An overview. Microstructure-Property Correlations for Hard, Superhard, and Ultrahard Materials, 1–23.

[11] Kimpel, R.F. & Moss, R.G. (1968). Melting point of 98.9 to 99.6% pure boron. Journal of Chemical & Engineering Data, 13(2), 231–234.

[12] Neuman, E.W., Hilmas, G.E. & Fahrenholtz, W.G. (2013). Strength of zirconium diboride to 2300 C. Journal of the American Ceramic Society, 96(1), 47–50.

[13] Neuman, E.W., Hilmas, G.E. & Fahrenholtz, W.G. (2016). Ultra-high temperature mechanical properties of a zirconium diboride–zirconium carbide ceramic. Journal of the American Ceramic Society, 99(2), 597–603.

[14] Hulm, J.K. & Matthias, B.T. (1951). New superconducting borides and nitrides. Physical Review, 82(2), 273.

[15] Gabani, S., Flachbart, K., Siemensmeyer, K. & Mori, T. (2020). Magnetism and superconductivity of rare earth borides. Journal of Alloys and Compounds, 821, 153201.

[16] Sevik, C., Bekaert, J., Petrov, M. & Milošević, M.V. (2022). High-temperature multigap superconductivity in two-dimensional metal borides. Physical Review Materials, 6(2), 024803.

[17] Katano, S., Fernandez-Baca, J.A., Funahashi, S., Mori, N., Ueda, Y. & Koga, K. (1993). Crystal structure and superconductivity of $La^{2-}xBaxCuO_4$ ($0.03 \leq x \leq 0.24$). Physica C: Superconductivity, 214(1–2), 64–72.

[18] Dzero, M., Xia, J., Galitski, V. & Coleman, P. (2016). Topological Kondo insulators. Annual Review of Condensed Matter Physics, 7, 249–280.

[19] Weng, H., Zhao, J., Wang, Z., Fang, Z. & Dai, X. (2014). Topological crystalline Kondo insulator in mixed valence ytterbium borides. Physical Review Letters, 112(1), 016403.

[20] Thalmeier, P., Akbari, A. & Shiina, R. (2021). Multipolar order and excitations in rare-earth boride Kondo systems. *In:* Rare-Earth Borides (pp. 615–690). Jenny Stanford Publishing.

[21] Li, L., Sun, K., Kurdak, C. & Allen, J.W. (2020). Emergent mystery in the Kondo insulator samarium hexaboride. Nature Reviews Physics, 2(9), 463–479.

[22] Werheit, H. (1988). Optical properties of the crystalline modifications of boron and boron-rich borides. Progress in Crystal Growth and Characterization, 16, 179–223.

[23] Labov, S., Bowyer, S. & Steele, G. (1985). Boron and silicon: Filters for the extreme ultraviolet. Applied Optics, 24(4), 576–578.

[24] Cline, J.P., Deslattes, R.D., Staudenmann, J.L., Kessler, E.G., Hudson, L.T., Henins, A. & Cheary, R.W. (2000). Certificate Srm 660a. NIST, Gaithersburg, MD, USA.

[25] Gild, J., Zhang, Y.Y., Harrington, T.J., Jiang, S.C., Hu, T., Quinn, M.C. et al. (2016). High-entropy metal diborides: A new class of high entropy materials and a new type of ultrahigh temperature ceramics, Scientific Reports, 6,37946.

[26] Zhang, Y., Jiang, Z.B., Sun, S.K., Guo, W.M., Chen, Q.S., Qiu, J.X. & Lin, H.T. (2019). Microstructure and mechanical properties of high-entropy borides derived from boro/carbothermal reduction. Journal of the European Ceramic Society, 39(13), 3920–3924.

[27] Gild, J., Wright, A., Quiambao-Tomko, K., Qin, M., Tomko, J.A., bin Hoque, M.S. & Luo, J. (2020). Thermal conductivity and hardness of three single-phase high-entropy metal diborides fabricated by borocarbothermal reduction and spark plasma sintering. Ceramics International, 46(5), 6906–6913.

[28] Tallarita, G., Licheri, R., Garroni, S., Orrù, R. & Cao, G. (2019). Novel processing route for the fabrication of bulk high-entropy metal diborides. Scripta Materialia, 158, 100–104.

[29] Qin, M., Gild, J., Wang, H., Harrington, T., Vecchio, K.S. & Luo, J. (2020). Dissolving and stabilizing soft WB2 and MoB2 phases into high-entropy borides via boron-metals reactive sintering to attain higher hardness. Journal of the European Ceramic Society, 40(12), 4348–4353.

[30] Feng, L., Fahrenholtz, W.G. & Hilmas, G.E. (2020). Processing of dense high-entropy boride ceramics. Journal of the European Ceramic Society, 40(12), 3815–3823.

[31] Qin, M., Yan, Q., Wang, H., Hu, C., Vecchio, K.S. & Luo, J. (2020). High-entropy monoborides: Towards superhard materials. Scripta Materialia, 189, 101–105.

[32] Pangilinan, L.E., Turner, C.L., Akopov, G., Anderson, M., Mohammadi, R. & Kaner, R.B. (2018). Superhard tungsten diboride-based solid solutions. Inorganic Chemistry, 57(24), 15305–15313.

[33] Qin, M., Yan, Q., Liu, Y. & Luo, J. (2021). A new class of high-entropy M3 B4 borides. Journal of Advanced Ceramics, 10, 166–172.

[34] Qin, M., Yan, Q., Liu, Y., Wang, H., Wang, C., Lei, T. & Luo, J. (2021). Bulk high-entropy hexaborides. Journal of the European Ceramic Society, 41(12), 5775–5781.

Nitrides and High-Entropy Nitrides

Increasingly, metal nitrides are serving as substitutes for traditional materials across a range of fields such as electrochemical systems [1], environmental cleanup [2], gas detection [3, 4], photocatalytic reactions [5], specialized ceramics [6, 7], and healthcare [8].

Various advanced materials, including composites, frequently incorporate transition metal nitride nanoparticles. Novel materials that are structured as nanoparticles and display various shapes—including hollow spheres, needles, capsules, particles, fibers, cubes, and plates—are the focus of considerable research attention [9-11]. This is due to their attributes like low mass density, thin structural walls, expansive surface areas, elevated toughness, adaptability, ionic conductivity, augmented chemical reactivity, and electrical characteristics.

In comparison to many unaltered metal oxides, metal nitrides offer advanced features such as high melting points, exceptional resistance to corrosion and wear, superior electronic properties, and effective catalytic characteristics [12]. Data on the hardness-to-melting point ratios (measured in Kelvin) for TiN, ZrN, VN, and NbN further substantiate their suitability for applications in high-stress and refractory environments [12].

Synthesis of Nitrides

In the evolving landscape of material science, numerous methods have been developed for the synthesis of metal nitrides. Currently employed techniques encompass traditional direct element bonding, solvothermal processes, carbothermal methods, sonochemical procedures, gas-phase synthesis, electrochemical methods, as well as solid-state metathesis and precursor methods involving solid-state metal-organic polymers and hydrazine sol-gel processes. Some of the most applied synthesis methods for metal nitrides are discussed in the following discussion.

Solid-state synthesis has traditionally been viewed as an efficient and environmentally responsible method for fabricating metal nitrides [13]. One notable benefit of this method is the tendency for the resultant products to be crystalline in structure.

Apart from conventional solid-state synthesis route, a recent review examines the fabrication of metal nitrides using the solid-solid separation method [12]. Here, ammonolysis of bulk ternary oxides is reported as performed under the flow of ammonia gas, at moderate temperatures ranging from 600°C to 800°C. This method is able to successfully produce various metal nitrides like TiN, NbN, VN, and Ta_3N_5, featuring large surface areas and pores ranging from approximately 10 to 40 nm. Such methodology allows for the formation of phase-pure, morphologically uniform products featuring surface areas ranging between 15.1 m²/g and 59.9 m²/g [14].

During the solid-state segregation method, reduction and nitridation occur simultaneously. For example, initial reduction-nitridation reactions begin with metal oxides, followed by a secondary nitridation step that leads to the formation of pure, single-phase metal nitrides.

The sol-gel synthesis method involves the formation of solid or nano-scale materials from a liquid solution via an intermediate gel stage [15]. In this technique, molecular-level mixing of reactants allows for quick reactions and the formation of highly homogeneous products. Specifically, the sol-gel process often employs metal alkoxides that are susceptible to nucleophilic attack, followed by hydrolysis and eventual condensation, leading to polymerization. This technique has been effective in synthesizing various types of nanoparticles, including metal nitrides [16-18].

The sol-gel approach, however, is complex and involves several stages of chemical processing. The rate of condensation can be manipulated by adding appropriate catalysts or polymeric networks to either slow down or speed up the process. Metal nitride synthesis via sol-gel often originates from metal-organic compounds, which are then transformed into final products through a series of reactions involving sodium acetylide and ammonia aqueous solution [19, 20].

The Solid-State (SS) metal oxide-organic technique is often cited as the most straightforward and cost-effective means of synthesizing both metal nitrides and carbides [12].

Within this approach, an initial precursor comprised of a metal oxide and an organic substance is reacted. This precursor is then transformed into the respective metal nitride and carbide at moderate heat levels ranging between 900 °C and 1100 °C. Compounds such as dicyandiamide, melamine, cyanamide, and ammonium cyanamide have been identified as effective agents for nitriding across varying temperatures [21-23]. Thus, this method is particularly versatile, suitable for synthesizing nitrides of most transition metals in Groups IV to VI.

Sintering of Nitrides

Typically, metal nitrides are generated in powder form and must be fashioned into specific geometries for practical use. The densification process usually consists of cold pressing the powders, also known as cold compaction, at pressures usually in the hundreds of megapascals. This is followed by a heat treatment at 0.5 to 0.6 times the material's melting point (T_{melt}), or alternatively, both steps may be combined in a process termed hot pressing.

Sintering—the transformation of these powder compacts into solid, dense ceramics upon heating—is integral in shaping the material's properties such as mechanical strength and dielectric response. The behavior of the powder particles during sintering influences the ultimate microstructure of a material, which is vital for defining its final properties. Elevated sintering temperatures (greater than 0.8 T_{melt}) increase the risk of oxidation and may promote lattice diffusion, leading to particle or grain coarsening [12]. Such coarsening can counteract the densification process, resulting in residual micro-porosity that may impact both structural and functional attributes of the material.

Recent advancements have seen the advent of alternative sintering techniques like Spark Plasma Sintering (SPS) and Microwave Sintering. These methods are increasingly being explored as replacements for traditional heat treatment and sintering approaches, especially for non-oxide ceramics like carbides and nitrides [24-26].

Applications of Nitrides

Metal nitrides, particularly those with high surface areas, play crucial roles in catalysis. They may function either as active catalysts or as supports for metal ions or particles with catalytic properties [27]. Applications are widespread and include oil refining [28], chemical manufacturing [29], and treatment of exhaust gases [30]. Just as a recent example, Ologunagba and Kattel [31] conducted Density Functional Theory (DFT) analyses to explore the hydrogen evolution reaction efficiency of core-shell Pt/Pd-monolayer-transition metal nitride catalysts, demonstrating that multiple transition metal nitrides (TMNs) exhibit both advantageous negative formation energies and low surface energies, making TMNs very promising catalysts.

Sensors based on metal nitrides have shown remarkable sensitivity and selectivity [12]. The electronic surface states of these materials enable this heightened sensing ability. An example can be drawn from the work of Qu et al. [32], that reported that a sensor made of Sn_3N_4 exhibited optimal sensitivity towards various analyte gases at a temperature as low as 120°C. Such sensors demonstrate a promising blend of sensitivity and selectivity, with particular effectiveness towards ethanol.

Metal nitrides have been researched also for their application in energy storage systems due to their thermal stability and resistance to stress [33, 34]. These properties are advantageous for long-lasting, efficient energy storage solutions.

Metal nitrides' high surface area and chemical stability make them excellent candidates for adsorbing and subsequently neutralizing various pollutants, thus being extensively considered for many environmental remediation applications [35, 36].

Finally, metal nitrides have notable mechanical strength and are resistant to thermal, chemical, and mechanical attacks. They can therefore replace metal oxides in specialized areas where these properties are required, such as aerospace or high-stress industrial environments [37].

However, the full potential of metal nitrides, especially in the form of nanoparticles, is yet to be completely realized. High-entropy metal nitrides represent one of the possible way to unlock such potential.

High-Entropy Nitrides

As already discussed, transition metal nitrides are prized for their exceptional hardness, resistance to wear, and high melting points—exceeding 1800 °C and in some cases reaching above 4000 °C. High-entropy variants of these materials offer enhanced hardness and thermal resilience, making them ideal for cutting-edge and thermally protective elements in aerospace engineering that demand bulk materials.

High-entropy nitrides (HENs) have firstly been fabricated in both powder and thin-film forms [38-40], due to their superior solubility and phase stability compared to bulk ceramics. For example, Jin et al. [ref] presented the formation of a specific powdered metal HEN using planetary ball milling followed by annealing. However, their work did not afford sintering or mechanical testing.

Conversely, Wen et al. [41] reported the first-ever bulk synthesis of a HEN composition, specifically (HfNbTaTiZr)N, and Dippo et al. [42] reported the creation of 5 additional bulk HENs.

In particular, Dippo's HENs exhibited a noteworthy improvement of the mechanical properties compared to the related mononitride precursors when evaluated using the rule-of-mixtures. Specifically, hardness escalated by an average of 22% in HENs. Concurrently, the elastic modulus values also saw an increase, rising on average by 17%.

These augmentations in mechanical properties are attributed to two primary factors: an upsurge in the configurational entropy and a reduction in the valence electron concentration. High configurational entropy is linked to superior strength and phase stability, whilst valence electron concentration (VEC) impacts the bond character of a metal nitride, considered to be a mixture of covalent, ionic, and metallic bonds. This mixed bond character contributes to the nitrides' unique features, such as maintaining high melting points and high thermal and electrical conductivities. By altering the material's composition, it is possible to manipulate the VEC, which in turn influences hardness and toughness. For instance, HENs with their elevated nitrogen content, and therefore higher VEC, could be expected to display lower hardness, suggesting possibilities for synthesizing ceramics with increased ductility in the future [42].

As for the HEN sintering, the work conducted by Moskovskikh et al. [43] demonstrates the successful application of SPS consolidation technology for producing high-quality bulk high-entropy nitrides. Moskovskikh's composition was inspired by $Ti_{0.5}Ta_{0.5}N$, a ternary ceramic with exceptionally high hardness and augmented with three additional metals (i.e. Hf, Nb and Zr) to achieve entropy stabilization while maintaining a Valence Electron Concentration (VEC) close to 9.5. The resulting material consisted mainly of hexagonal HEN grains and exhibited remarkable mechanical properties, such as hardness and fracture toughness, exceeding theoretical estimates and comparable materials. Even in this case, the enhancement in mechanical properties is believed to result from both entropy stabilization and an optimized VEC.

References

[1] Balogun, M.S., Huang, Y., Qiu, W., Yang, H., Ji, H. & Tong, Y. (2017). Updates on the development of nanostructured transition metal nitrides for electrochemical energy storage and water splitting. Materials Today, 20(8), 425–451.

[2] Kumar, A., Thakur, P.R., Sharma, G., Naushad, M., Rana, A., Mola, G.T. & Stadler, F.J. (2019). Carbon nitride, metal nitrides, phosphides, chalcogenides, perovskites and carbides nanophotocatalysts for environmental applications. Environmental Chemistry Letters, 17, 655–682.

[3] Schalwig, J., Müller, G., Ambacher, O. & Stutzmann, M. (2001). Group-III-nitride based gas sensing devices. Physica Status Solidi (A), 185(1), 39–45.

[4] Sharma, N., Pandey, V., Gupta, A., Tan, S.T., Tripathy, S. & Kumar, M. (2022). Recent progress on group III nitride nanostructure-based gas sensors. Journal of Materials Chemistry C, 10(34), 12157–12190.

[5] Cheng, Z., Qi, W., Pang, C.H., Thomas, T., Wu, T., Liu, S. & Yang, M. (2021). Recent advances in transition metal nitride-based materials for photocatalytic applications. Advanced Functional Materials, 31(26), 2100553.

[6] Bendriss, K., Rached, H., Ouadha, I., Azzouz-Rached, A., Chahed, A., Bentouaf, A., ... & Rached, D. (2023). The stability, mechanical, electronic, and thermal features of the new superhard double transition-metal mono-nitrides and mono-carbides compounds. Indian Journal of Physics, 97(4), 1125–1135.

[7] Pogrebnjak, A., Smyrnova, K. & Bondar, O. (2019). Nanocomposite multilayer binary nitride coatings based on transition and refractory metals: Structure and properties. Coatings, 9(3), 155.

[8] Heimann, R.B. (2021). Silicon nitride, a close to ideal ceramic material for medical application. Ceramics, 4(2), 208–223.

[9] Dongil, A.B. (2019). Recent progress on transition metal nitrides nanoparticles as heterogeneous catalysts. Nanomaterials, 9(8), 1111.

[10] He, T., Wang, Z., Li, X., Tan, Y., Liu, Y., Kong, L., ... & Ran, F. (2019). Intercalation structure of vanadium nitride nanoparticles growing on graphene surface toward high negative active material for supercapacitor utilization. Journal of Alloys and Compounds, 781, 1054–1058.

[11] Wu, H., Jiang, H., Yang, Y., Hou, C., Zhao, H., Xiao, R. & Wang, H. (2020). Cobalt nitride nanoparticle coated hollow carbon spheres with nitrogen vacancies as an electrocatalyst for lithium–sulfur batteries. Journal of Materials Chemistry A, 8(29), 14498–14505.

[12] Ashraf, I., Rizwan, S. & Iqbal, M. (2020). A comprehensive review on the synthesis and energy applications of nano-structured metal nitrides. Frontiers in Materials, 7, 181.

[13] Aykol, M., Montoya, J.H. & Hummelshøj, J. (2021). Rational solid-state synthesis routes for inorganic materials. Journal of the American Chemical Society, 143(24), 9244–9259.

[14] Yang, M., MacLeod, M.J., Tessier, F. & DiSalvo, F.J. (2012). Mesoporous metal nitride materials prepared from bulk oxides. Journal of the American Ceramic Society, 95, 3084–3089.

[15] Bokov, D., Turki Jalil, A., Chupradit, S., Suksatan, W., Javed Ansari, M., Shewael, I.H. & Kianfar, E. (2021). Nanomaterial by sol-gel method: Synthesis and application. Advances in Materials Science and Engineering, 2021, 1–21.

[16] Barshilia, H.C., Lakshmi, R.V., Bera, P., Basu, B.B.J., Srinivas, G., Kumar, V.P. & Manikandanath, N.T. (2022). Transition metal nitride/oxide-based multilayer PVD coating with sol-gel derived ormosil passivation layer as an efficient solar absorber: Studies on high temperature stability and performance evaluation. Solar Energy, 239, 283–293.

[17] Kayani, Z.N., Bashir, Z., Mohsin, M., Riaz, S. & Naseem, S. (2021). Sol-gel synthesized boron nitride (BN) thin films for antibacterial and magnetic applications. Optik, 243, 167502.

[18] Sheng, Y., Yang, J., Wang, F., Liu, L., Liu, H., Yan, C. & Guo, Z. (2019). Sol-gel synthesized hexagonal boron nitride/titania nanocomposites with enhanced photocatalytic activity. Applied Surface Science, 465, 154–163.

[19] Giordano, C. &Antonietti, M. (2011). Synthesis of crystalline metal nitride and metal carbide nanostructures by sol-gel chemistry. Nano Today, 6, 366–380.

[20] Baxter, D.V., Chisholm, M.H., Gama, G.J., DiStasi, V.F., Hector, A.L. & Parkin, I.P. (1996). Molecular routes to metal carbides, nitrides, and oxides. 2. Studies of the ammonolysis of metal dialkylamides and hexamethyldisilylamides. Chemistry of Materials, 8, 1222–1228.

[21] Jürgens, B., Irran, E., Senker, J., Kroll, P., Müller, H. &Schnick, W. (2003). Melem (2, 5, 8-triamino-tri-s-triazine), an important intermediate during condensation of melamine rings to graphitic carbon nitride: Synthesis, structure determination by X-ray powder diffractometry, solid-state NMR, and theoretical studies. Journal of the American Chemical Society, 2003, 10288–10300.

[22] Lei, M., Zhao, H., Yang, H., Li, P., Tang, H., Song, B. et al. (2007). Ammonium dicyanamide as a precursor for the synthesis of metal nitride and carbide nanoparticles. Diamond and Related Materials, 16, 1974–1981.

[23] Koziej, D., Krumeich, F., Nesper, R. & Niederberger, M. (2009). Nonaqueous liquid-phase synthesis of nanocrystalline metal carbodiimides. A proof of concept for copper and manganese carbodiimides. Journal of Materials Chemistry, 19, 5122–5124.

[24] Liang, L., Wei, B., Wang, D., Fang, W., Chen, L. & Wang, Y. (2022). Densification, microstructures, and mechanical properties of (Zr, Ti)(C, N) ceramics fabricated by spark plasma sintering. Journal of the European Ceramic Society, 42(14), 6445–6456.

[25] Chávez, J.M., Moshtaghioun, B.M., Hernández, F.C. & García, D.G. (2019). Sintering kinetics, defect chemistry and room-temperature mechanical properties of titanium nitride prepared by spark plasma sintering. Journal of Alloys and Compounds, 807, 151666.

[26] Li, W., Li, W., Chen, J., Ying, Y., Yu, J., Zheng, J. & Che, S. (2022). Migration of N element and evolution of microstructure in spark plasma sintered bulk γ'-Fe4N. Journal of Alloys and Compounds, 928, 167201.

[27] Dongil, A.B. (2019). Recent progress on transition metal nitrides nanoparticles as heterogeneous catalysts. Nanomaterials, 9(8), 1111.

[28] Ullah, S., Khan, K., Farooq, M.U., Ahmad, W., Naeem, M., Subhan, F. & Yaseen, M. (2022). Perspectives on advances in the catalytic desulfurization and denitrogenation of transportation fuel oils using graphitic carbon nitride and boron nitride. Energy & Fuels, 36(16), 8900–8924.

[29] Akhundi, A., Badiei, A., Ziarani, G.M., Habibi-Yangjeh, A., Munoz-Batista, M.J. & Luque, R. (2020). Graphitic carbon nitride-based photocatalysts: Toward efficient organic transformation for value-added chemicals production. Molecular Catalysis, 488, 110902.

[30] Liu, D., Yang, L., Ling, Y., Wu, J., Li, B. & Li, C. (2022). Graphitic carbon nitride for gaseous mercury emission control: A review. Energy & Fuels, 36(8), 4297–4313.

[31] Ologunagba, D. & Kattel, S. (2022). Pt-and Pd-modified transition metal nitride catalysts for the hydrogen evolution reaction. Physical Chemistry Chemical Physics, 24(20), 12149–12157.

[32] Qu, F., Yuan, Y. & Yang, M. (2017). Designed synthesis of Sn_3N_4 nanoparticles through soft urea route with excellent gas sensing properties. Chemistry of Materials, 29(3), 969–974.

[33] Raveh, A., Zukerman, I., Shneck, R., Avni, R. & Fried, I. (2007). Thermal stability of nanostructured superhard coatings: A review. Surface and Coatings Technology, 201(13), 6136–6142.

[34] Kumar, D.D., Kumar, N., Kalaiselvam, S., Dash, S. & Jayavel, R. (2017). Wear resistant super-hard multilayer transition metal-nitride coatings. Surfaces and Interfaces, 7, 74–82.

[35] Aissani, L., Alhussein, A., Zia, A.W., Mamba, G. & Rtimi, S. (2022). Magnetron sputtering of transition metal nitride thin films for environmental remediation. Coatings, 12(11), 1746.

[36] Kumar, S., Reddy, K.R., Reddy, C.V., Shetti, N.P., Sadhu, V., Shankar, M.V., ... & Aminabhavi, T.M. (2021). Metal nitrides and graphitic carbon nitrides as novel photocatalysts for hydrogen production and environmental remediation. Nanostructured Materials for Environmental Applications, 485–519.

[37] Xavier, J.R. & Jeeva, N. (2022). Evaluation of newly synthesized nanocomposites containing thiazole modified aluminium nitride nanoparticles for aerospace applications. Materials Chemistry and Physics, 286, 126200.

[38] Jin, T., Sang, X., Unocic, R.R., Kinch, R.T., Liu, X., Hu, J. & Dai, S. (2018). Mechanochemical-assisted synthesis of high-entropy metal nitride via a soft urea strategy. Advanced Materials, 30(23), 1707512.

[39] Xing, Q.W., Xia, S.Q., Yan, X.H. & Zhang, Y. (2018). Mechanical properties and thermal stability of (NbTiAlSiZr)Nx high-entropy ceramic films at high temperatures. Journal of Materials Research, 33(19), 3347–3354.

[40] Li, H., Jiang, N., Li, J., Huang, J., Kong, J. & Xiong, D. (2021). Hard and tough (NbTaMoW)Nx high entropy nitride films with sub-stoichiometric nitrogen. Journal of Alloys and Compounds, 889, 161713.

[41] Wen, T., Ye, B., Nguyen, M.C., Ma, M. & Chu, Y. (2020). Thermophysical and mechanical properties of novel high-entropy metal nitride-carbides. Journal of the American Ceramic Society, 103(11), 6475–6489.

[42] Dippo, O.F., Mesgarzadeh, N., Harrington, T.J., Schrader, G.D. & Vecchio, K.S. (2020). Bulk high-entropy nitrides and carbonitrides. Scientific Reports, 10(1), 21288.

[43] Moskovskikh, D., Vorotilo, S., Buinevich, V., Sedegov, A., Kuskov, K., Khort, A., ... & Mukasyan, A. (2020). Extremely hard and tough high entropy nitride ceramics. Scientific Reports, 10(1), 19874.

Silicides and High-Entropy Silicides

Silicides are binary mixtures composed of silicon and other elements with higher electronegativity. The nature of the chemical bonds in silicides can range from primarily covalent to primarily ionic, contingent upon the electronegativities of the elements involved. This makes them analogous to borides and carbides, as their composition does not conform neatly to covalent molecular structures. Chemical bonding in silicides spans from metal-like configurations to covalent or ionic bonding [1]. The metal-like attributes often confer electrical conductivity to many silicides. Specifically, transition metal silicides are generally resistant to aqueous solutions, with hydrofluoric acid being a notable exception.

Generally speaking, silicides are hard, crystalline compounds, possessing some of the mechanical properties of metals as well as many of their physical and chemical features. Silicides do not occur naturally and must be artificially synthesized. One common method for their synthesis is to deposit a refractory metal directly onto a silicon surface, thereby triggering a reaction that forms the desired silicide layer. After deposition, the metal-silicon system is subjected to elevated temperatures to facilitate the chemical reactions between the constituents. Typically, silicides with higher metal content are formed initially and continue to grow until the metal is fully consumed. Once this occurs, silicides with lower metal concentrations begin to emerge, further growing by utilizing the metal-rich silicides. Ultimately, the system stabilizes, often forming a bisilicide $MeSi_2$, where 'Me' represents a generic metal.

The majority of metals can react with silicon to form a metal silicide, yet not all resulting silicides hold value for practical engineering or structural purposes [2]. For high-temperature applications, the silicides of interest are predominantly those derived from refractory metals in groups IVA, VA, and VIA of the periodic table. This includes silicides of titanium, zirconium, hafnium, vanadium, niobium, tantalum, chromium, molybdenum, and tungsten [3-6]. Conversely, in the realm of microelectronics, silicides containing group VIIIA metals have great technological interest [7]. They are commonly used for contacts, gates, and interconnections in silicon-based integrated circuits. Among these, $FeSi_2$, $CoSi_2$, $NiSi_2$, and $PtSi$ are particularly noteworthy, primarily due to their low electrical resistivity [8-10].

CoSi$_2$ is notably the most interesting metal silicide, largely owing to its low electrical resistivity, approximately 14 $\mu\Omega \cdot$ cm [2], and its crystalline structure that closely resembles that of silicon [11]. Cobalt bisilicide exhibits a CaF$_2$-type symmetry with a lattice parameter of a = 5.365 Å.

NiSi$_2$ is another cubic metal bisilicide of significant interest, especially within the field of Ni–Si alloys for device applications [12, 13]. Although the nickel–silicon system features multiple phases, only three primarily govern the silicidation process: Ni$_2$Si owning a resistivity of 24 $\mu\Omega \cdot$ cm, NiSi a resistivity of 14 $\mu\Omega \cdot$cm, and NiSi$_2$ with a resistivity of 34 $\mu\Omega \cdot$cm [1].

Germanium silicide (often referred to as silicon germanium or SiGe) is the most interesting among non-metal silicides for its peculiar thermoelectric properties [14]. Particularly, at elevated temperatures (around 1300 K), it exhibits a large Seebeck coefficient (>200 μV/K) and low thermal conductivity (<5 W/m^{-1}K^{-1}), attributed to its relatively broad band gap (~0.9 eV). These properties make it well-suited for high-temperature power generation uses [7, 15].

Silicides Fabrication

In terms of production, silicides are employed in both thin film forms for device fabrication and in bulk forms for other specific applications.

Several methodologies, including arc casting and directional solidification, have been implemented to produce bulk silicides, with the arc casting being the most employed one [2].

In particular, arc casting can be executed through either consumable or non-consumable vacuum-arc melting techniques. This approach is suitable for producing high-melting silicides like TiSi$_2$ and MoSi$_2$ [16].

A range of thin film coating methodologies exists, broadly falling into two primary categories: chemical vapor deposition (CVD) and physical vapor deposition (PVD). Both approaches offer the flexibility to execute the process through either the deposition of a single constituent or the co-deposition of multiple constituents (also referred to as simultaneous deposition), as in the case of metal silicides [17, 18].

Silicides Applications

Many metal silicides possess melting temperatures and band gaps suitable for medium-temperature power generation applications. Notably, CoSi$_2$, Mg$_2$Si, MnSi$_x$, CrSi$_2$, β-FeSi$_2$ are among the most promising [7]. Mg$_2$Si and its derived doped compounds have attracted recent scholarly attention due to their high thermoelectric performance in a medium-temperature range, comparable to other effective thermoelectric materials like lead telluride [19, 20]. Their natural abundance and non-toxicity further augment their appeal, driving research to improve their thermoelectric properties using methods such as nanostructuring.

Nanostructuring has already demonstrated improvements for germanium silicides [21, 22].

In semiconductor technology, silicides are extensively used for low-resistivity contacts and interconnects. The term "germanosilicide" applies to compounds formed between SiGe and metals like Ti, Co, Ni, and Pt [7]. For high-temperature applications, silicon germanium ($Si_{1-x}Ge_x$) and ruthenium silicide (Ru_2Si_3) are strong contenders. Nanostructured $Si_{1-x}Ge_x$ has indicated enhanced thermoelectric properties for both p- and n-type materials. Conversely, Ru_2Si_3 also shows promise in thermoelectric applications and could potentially outperform $Si_{1-x}Ge_x$ at high temperatures if optimum carrier concentration could be achieved, which has been historically challenging to accomplish [7].

High-Entropy Silicides

In 2019, Gild et al. [23] were the first to successfully create a high-entropy metal bisilicide (HES), specifically $(Mo_{0.2}Nb_{0.2}Ta_{0.2}Ti_{0.2}W_{0.2})Si_2$. Such unique compound features a hexagonal C40 crystal arrangement with an ABC stacking sequence, belonging to the P6222 space group. Gild's HES exhibits noteworthy mechanical properties, with a nanohardness of around 17 GPa and a Vickers hardness of around 12 GPa. Additionally, it possesses a thermal conductivity significantly lower—by an order of magnitude—compared to the commonly utilized tetragonal $MoSi_2$, and roughly one-third of the values reported for hexagonal $NbSi_2$ and $TaSi_2$, which share the same crystal structure.

The exploration of High-Entropy Silicides (HES) holds particular promise for applications in the field of microelectronics, where silicides are commonly used, notably in the insulation of metal-oxide-semiconductor field-effect-transistors (MOSFETs). However, still few research studies reported stable HES compositions [24-26] and very few compositions achieved a single-phase material, containing oxides or intermetallics as secondary phases, either due to processing conditions or thermodynamic stability. This evidence highlights the need for deeper insights into phase formation and evolution of HES materials.

A useful contribution to address these challenges comes from Vyatskikh et al. [27], who adopted the CALculation of PHAse Diagrams (CALPHAD) approach, supported by advancements in computational thermodynamics databases, thus offering a reliable tool for predicting phase stability and identifying promising candidate HES compositions. In their study, Vyatskikh et al. utilized CALPHAD to pinpoint two single-phase (with a C40 hexagonal crystal structure) HES materials, i.e. the ternary $(CrMoTa)Si_2$ and the quinary $(Cr_{0.2}Mo_{0.2}Ta_{0.2}V_{0.2}Nb_{0.2})Si_2$, successively synthesizing them in thin film form through electron beam evaporation onto a silicon substrate, followed by a heat treatment that triggered a solid-state reaction.

Noteworthy, in the research conducted by Wen et al. [28], a novel category of high-entropy materials referred to as high-entropy alumino-silicides (HEAS) was introduced, as a derivation of HES. Specifically, they introduced a HEAS with the following composition: $(Mo_{0.25}Nb_{0.25}Ta_{0.25}V_{0.25})(Al_{0.5}Si_{0.5})_2$, featuring a unique multi-cationic and multi-anionic structural configuration. Firstly, Wen et al. performed an initial theoretical scrutiny by using first-principles calculations, focusing on thermodynamics and lattice size differences of the various involved elements; subsequently, they successfully fabricated the $(Mo_{0.25}Nb_{0.25}Ta_{0.25}V_{0.25})(Al_{0.5}Si_{0.5})_2$ system through a solid-state reaction at 1573 K. Wen's HEAS demonstrated a homogenous single-crystal hexagonal structure characteristic of metal alumino-silicides, along with a high degree of compositional uniformity.

Even though both HES and HEAS research studies are at their infancy, such high-entropy compounds pave the way for future complex multi-cationic and multi-anionic compounds to be used in microelectronics and thermoelectric applications [29].

References

[1] Wu, Z.J. & Su, Z.M. (2006). Electronic structures and chemical bonding in transition metal monosilicides MSi (M = 3d, 4d, 5d elements). The Journal of Chemical Physics, 124(18).

[2] Pelleg, J. (2019). Mechanical Properties of Silicon Based Compounds: Silicides. Springer.

[3] Paul, A. (2019). Diffusion rates of components in metal-silicides depending on atomic number of refractory metal component. Diffusion Foundations, 21, 29–84.

[4] Wu, Y., Yang, C. & Duan, Y. (2021). First-principles exploration of elastic anisotropy and thermal properties of the C40-type VSi_2, $NbSi_2$, and $TaSi_2$ disilicides. Materials Today Communications, 29, 102818.

[5] Tang, P., Hu, C., Pang, S., Jiang, Y., Liang, B., Wang, L. & Tang, S. (2023). Self-densification behavior, interfacial bonding and cyclic ablation resistance of $HfSi_2$-$ZrSi_2$ modified SiC/ZrB_2-SiC/SiC coating for Cf/SiC composite. Corrosion Science, 219, 111223.

[6] Pan, Y., Jing, C. & Wu, Y.P. (2019). The structure, mechanical and electronic properties of WSi_2 from first-principles investigations. Vacuum, 167, 374–381.

[7] Nozariasbmarz, A., Agarwal, A., Coutant, Z.A., Hall, M.J., Liu, J., Liu, R. & Vashaee, D. (2017). Thermoelectric silicides: A review. Japanese Journal of Applied Physics, 56(5S1), 05DA04.

[8] Abbassi, L., Mesguich, D., Berthebaud, D., Le Tonquesse, S., Srinivasan, B., Mori, T. & Beaudhuin, M. (2021). Effect of Nanostructuring on the Thermoelectric Properties of β-$FeSi_2$. Nanomaterials, 11(11), 2852.

[9] Wolansky, D., Blaschke, J.P., Drews, J., Grabolla, T., Heinemann, B., Lenke, T., ... & Deyo, D. (2020). Nickel and nickel-platinum silicide for BiCMOS devices. ECS Transactions, 98(5), 351.

[10] Chiu, S.P., Tsuei, C.C., Yeh, S.S., Zhang, F.C., Kirchner, S. & Lin, J.J. (2021). Observation of triplet superconductivity in $CoSi_2/TiSi_2$ heterostructures. Science Advances, 7(29), eabg6569.

[11] Vartanian, A. (2023). Sneaking single metal atoms into silicon. Nature Reviews Materials, 1–1.

[12] Umirzakov, B.E., Tashmukhamedova, D.A., Tashatov, A.K. & Mustafoeva, N.M. (2019). Electronic and optical properties of $NiSi_2$/Si nanofilms. Technical Physics, 64, 708–710.

[13] Pandey, R.K., Maity, G., Pathak, S., Kalita, P. & Dubey, S. (2022). New insights on Ni-Si system for microelectronics applications. Microelectronic Engineering, 111871.

[14] Borisenko, V.E. (2013). Semiconducting Silicides: Basics, Formation, Properties (Vol. 39). Springer Science & Business Media.

[15] Wang, W., Yuan, B. & Zhou, C. (2014). Formation and oxidation resistance of germanium modified silicide coating on Nb based in situ composites. Corrosion Science, 80, 164–168.

[16] Boettinger, W.J., Perepezko, J.H. & Frankwicz, P.S. (1992). Application of ternary phase diagrams to the development of $MoSi_2$-based materials. Materials Science and Engineering: A, 155(1–2), 33–44.

[17] Hamzan, N., Sookhakian, M., Mohd Sarjidan, M.A., Tripathi, M., Dalton, A.B., Goh, B.T. & Alias, Y. (2023). Synthesis of manganese silicide nanowires by thermal chemical vapor deposition for the hydrogen evolution reaction. ACS Applied Nano Materials, 6(13), 12140–12149.

[18] Huang, W., Srot, V., Wagner, J. & Richter, G. (2019). Fabrication of α-$FeSi_2$ nanowhiskers and nanoblades via electron beam physical vapor deposition. Materials & Design, 182, 108098.

[19] Tani, J.I. & Kido, H. (2005). Thermoelectric properties of Bi-doped Mg_2Si semiconductors. Physica B: Condensed Matter, 364(1–4), 218–224.

[20] Seth, P.P., Parkash, O. & Kumar, D. (2020). Structure and mechanical behavior of in situ developed Mg_2 Si phase in magnesium and aluminum alloys – A review. RSC Advances, 10(61), 37327–37345.

[21] Yuhara, J., Shimazu, H., Kobayashi, M., Ohta, A., Miyazaki, S., Takakura, S.I., ... & Le Lay, G. (2021). Epitaxial growth of massively parallel germanium nanoribbons by segregation through Ag (1 1 0) thin films on Ge (1 1 0). Applied Surface Science, 550, 149236.

[22] Chiang, P.T., Hu, S.M., Yen, W.T., Wu, H.J., Hsu, H.P. & Lan, C.W. (2023). A study of iron-doped SiGe growth for thermoelectric applications. Journal of Alloys and Compounds, 171700.

[23] Gild, J., Braun, J., Kaufmann, K., Marin, E., Harrington, T., Hopkins, P., ... & Luo, J. (2019). A high-entropy silicide: $(Mo_{0.2}Nb_{0.2}Ta_{0.2}Ti_{0.2}W_{0.2})Si_2$. Journal of Materiomics, 5(3), 337–343.

[24] Qin, Y., Liu, J.X., Li, F., Wei, X., Wu, H. & Zhang, G.J. (2019). A high entropy silicide by reactive spark plasma sintering. Journal of Advanced Ceramics, 8, 148–152.

[25] Liu, D., Huang, Y., Liu, L. & Zhang, L. (2020). A novel of MSi_2 high-entropy silicide: Be expected to improve mechanical properties of $MoSi_2$. Materials Letters, 268, 127629.

[26] Qin, Y., Wang, J.C., Liu, J.X., Wei, X.F., Li, F., Zhang, G.J., ... & Wu, H. (2020). High-entropy silicide ceramics developed from (TiZrNbMoW) Si_2 formulation doped with aluminum. Journal of the European Ceramic Society, 40(8), 2752–2759.

[27] Vyatskikh, A.L., MacDonald, B.E., Dupuy, A.D., Lavernia, E.J., Schoenung, J.M. & Hahn, H. (2021). High entropy silicides: CALPHAD-guided prediction and thin film fabrication. Scripta Materialia, 201, 113914.

[28] Wen, T., Liu, H., Ye, B., Liu, D. & Chu, Y. (2020). High-entropy alumino-silicides: A novel class of high-entropy ceramics. Science China Materials, 63(2), 300–306.

[29] Salian, A., Sengupta, P., Vishalakshi Aswath, I., Gowda, A. & Mandal, S. (2023). A review on high entropy silicides and silicates: Fundamental aspects, synthesis, properties. International Journal of Applied Ceramic Technology, 20(5), 2635–2660.

Empirical Descriptors and High-Entropy Materials Design Rules

Critical Factors affecting High-entropy Single-phase Stability

In applying Boltzmann's equation $(\Delta S_{conf} = -R\sum_{i}^{N} x_i \ln(x_i))$ for an ideal chemical system, one must consider that there will invariably be a standard deviation in entropy due to imperfect boundary conditions. To accurately describe the formation of a single-phase system and counteract its non-ideality, a statistical methodology is essential, taking into account the imperfect influences of thermodynamics, kinetics, and electric fields across the system.

The weighted average of a specific entity being examined serves as the correction factor for subsequent calculations. For a general entity ϕ, the weighted average can be calculated as follows:

$$\phi = \frac{\sum_{i}^{N} c_i \phi_i}{N} \tag{1}$$

Here, N represents the total number of components in the system ($N \geq 5$), and c_i denotes the concentration of each element within the system.

As explained in detail in Chapter 4, a change in configurational entropy (ΔS_{conf}) influences the entropic component of the Gibbs-Helmholtz equation:

$$\Delta G_{mix} = \Delta H_{mix} - T\Delta S_{mix} = \sum_{i \neq j=1} \beta_{i,j} x_i x_j - T\Delta S_{mix} \tag{2}$$

where $b_{i,j}$ is the molar constant specific to the involved species, and x_i and x_j represent their mole fractions. Therefore, the enthalpic term's magnitude is a crucial factor to assess the stability of a high-entropy material (HEM) at a certain temperature.

Recently, Zhang et al. [1] have performed an experimental investigation to estimate the enthalpy of mixing for multi-component systems (i.e. possible HEMs) based on Miedema's model [2]. Their proposed equation for ΔH_{mix} for such system is:

$$\Delta H_{mix} = \sum_{j \neq i = 1}^{N} 4 \Delta H_{AB}^{mix} c_i c_j \tag{3}$$

Subsequent studies following Zhang's research [3] have concluded that for a non-ideal solution, a single-phase solid solution occurs only when the mixing enthalpy is moderately positive (up to 7 kJ/mol) or not overly negative (-49 kJ/mol $\leq \Delta H_{mix} \leq -5.5$ kJ/mol). In other words, for a stable multicomponent single-phase solution to occur, the absolute value of the mixing enthalpy should be as close to zero as possible.

The second key factor to assess the stability of a high-entropy material (HEM) is the atomic size difference (δ) between each component of the given system. A possible equation for calculating phase stability based on atomic size difference is:

$$\delta = 100 \sqrt{\sum_{i}^{N \geq 5} c_i \left(1 - \frac{r_i}{r} \right)^2} \tag{4}$$

Here, r_i is the atomic radius of each component of the system, and r is the average atomic radius.

The third key aspect to assess the stability of a high-entropy material (HEM) focuses on the electronegativity of the individual elements in the given multicomponent system. According to Pauling's rules, the formation of ionic compounds depends on various factors including coordination polyhedron of anions, ion sizes, and electrostatic bonds. Critical radius ratios and their corresponding coordination numbers can be used to determine the polyhedron type around a central ion. Additional findings suggest that the number of different types of constituents in a crystal is generally limited ("rule of parity"), and more straightforward structures are often favored.

The deviation in electronegativity for a multi-component system from the standard condition is addressed through a factor known as the average electronegativity of n-components. Using Pauling's general theory, a single solid phase requires a specific $\Delta\chi$, calculated as:

$$\Delta\chi = \sqrt{\sum_{i=1}^{n} c_i \left(\chi - \chi_i \right)^2} \tag{5}$$

Alternate theories like the Allen electronegativity theory proposed by Poletti et al. [4] quantify a range ($3 < D\chi < 6$) within which a stable multicomponent single-phase alloy can be formed.

The fourth key metric to take into account to assess the stability of a high-entropy material (HEM) is the "Concentration of Valence Electrons" (VEC), indicating the total count of electrons in an atom's outermost shell.

For a generic multi-component system, VEC can be calculated using the formula:

$$VEC = \sum_{i}^{N} c_i VEC_i \tag{6}$$

Chen et al. [5] previously attempted to forecast the obtained single-phase structure in High-Entropy Alloys (HEAs) via VEC postulating the ensuing conditions:

VEC $\geq$ 8 leads to a Face-Centered Cubic (FCC) structure.

$8 \leq$ VEC $\geq$ 6.87 results in either FCC or Body-Centered Cubic (BCC) structures.

VEC $<$ 6.87 results in a BCC structure.

Subsequent researches [6, 7] for multicomponent systems featuring heavy lanthanide group elements, a Hexagonal Close-Packed (HCP) structure is more likely to form, provided that VEC $\geq$ 3.

A fifth factor to be taken into account for High-Entropy Material (HEM) design, initially defined in the context of HEAs, is the individual melting points of the constituent elements. Yang et al. [8] proposed an index ω, defined as:

$$\omega = \frac{T_{i\_melt} \Delta S_{mix}}{\Delta H_{mix}} \geq 1 \tag{7}$$

In real-world scenarios, the equimolarity of the mixture is not perfect, affecting all previously discussed parameters. Thus, the total mixing entropy is a function of both the ideal configurational entropy (S_{id_conf}) and a term that accounts for the actual conditions:

$$S_{tot\_conf}\left(c_t, r_t, \xi\right) = S_{id\_conf}\left(c_{id}\right) + S_{real\_conf}\left(c_r, r_r, \xi\right) \tag{8}$$

Here, S_{real_conf} is influenced by the elemental concentrations (c_t), their atomic radii (r_t), and packing densities (ξ). The precondition for forming a single-phase system, taking into account the real configurational entropy, is:

$$S_{real_{conf}}\left(c_r, r_r, \xi\right) \ll S_{id_{conf}}\left(c_{id}\right) - \frac{\left|H_{mix}\right|}{T} \tag{9}$$

The larger the disparity between the ideal and actual configurational entropies, the less likely it is to find a single phase within a given multicomponent system. To quantify the excess of mixing entropy's impact on the crystal structure, Ye et al. [9] defined a parameter ϕ as:

$$\phi = \frac{S_{id_{conf}}\left(c_{id}\right) - \left|H_{mix}\right|}{T_{id\_melt} S_{real_{conf}}\left(c_r, r_r, \xi\right)} \tag{10}$$

Such parameter provides a measure to understand how deviations in mixing entropy can affect the likelihood of forming a single-phase in a multicomponent system (favored when $\phi > 20$).

The Cluster-plus-Glue Atom Model Approach

Understanding the crystal structures and chemical properties of High-Entropy Materials (HEMs) fundamentally relies on electron interactions within the material. HEMs present a complex scenario, as their single-phase formations are dictated by localized chemical orders among their constituting elements. This localized ordering complicates descriptions of the interactions among primary components. In such materials, mixing enthalpies of systems involving five or more elements can produce local chemical orders, which in turn make it more challenging for dislocations to move through the lattice. These enthalpic effects can result in local regions of order or disorder, best understood through statistical modeling.

To tackle this complexity, a model founded on the concept of cluster polyhedrons, i.e. the Cluster-plus-Glue Atom Model [10, 11] has been recently applied in the field of HEMs [12]. This model describes a crystal lattice in terms of a 'nearest-neighbor cluster,' identified as the most stable among the possible clusters (also called the 'principal cluster'), and one or more external 'glue atoms'. The Cluster-plus-Glue Atom Model relies on radial atomic density distribution ρ_a, dividing the total number of atoms by the spherical volume within a certain radial distance to define the principal cluster. In other words, the primary cluster is the one possessing the maximum atomic density relative to other possible structures.

To be clearer for the readership, the fluorite crystal structure will be taken as an example for defining cluster units and glue atoms.

Specifically, the structure of fluorite (CaF_2) is characterized by two dissimilar atomic sites: calcium cations occupying Wyckoff position 4a, and fluorine anions located in Wyckoff position 8c. Thus, fluorite structure could be described by using two distinct clusters: one with a local coordination number (CN) of 8 in the form of $[Ca–F_8]$ and another one with a CN = 10 as $[F–Ca_4F_6]$. The concept of local coordination number is as outlined by Miracle [13] in his model for the dense packing of atomic clusters in metallic glasses. The ionic radii of Ca and F are 0.112 nm (when in eightfold coordination) and 0.131 nm (when in fourfold coordination), respectively. Given these dimensions, the $[Ca–F_8]$ cluster can be designed as the principal cluster since it exhibits the highest density of atomic packing. Accordingly, the 'glue atoms' are those that are situated at the second neighbors or farther from the principal clusters.

For a generic inorganic compound, the smallest cluster formula according to this model consists of the principal cluster (or the smallest number of co-principal clusters for more complex systems), supplemented by the minimum quantity of 'glue atoms' necessary to meet the chemical octet rule.

Therefore, the most concise cluster formula for the fluorite structure, according to the Cluster-plus-Glue Atom Model, is $[Ca–F_8]Ca_3$. Here, $[Ca–F_8]$ serves as the principal cluster and Ca_3 represents the 'glue atoms'. The e/u ratio (number of valence electrons per molecular formula) for $[Ca–F_8]Ca_3$ is 64, i.e. a multiple of 8, in order to comply with the octet rule. Definitely, the Cluster-plus-Glue formula of a given structure can be viewed as its molecular-like building block.

The energy bonding (E) of particles within the principal cluster plays a significant role in defining the overall cluster density. As Du et al. proposed recently [14], the intra-cluster energy interactions exceed those required to maintain a single atom's attachment to the cluster. They introduced a parameter, IFC, to describe the force exerted on an atom due to another atom's displacement. For a system composed by N atoms, such force can be quantified as follows:

$$ICF = \sqrt{\left(\frac{\delta^2 E}{\delta x^2}\right)^2 + \left(\frac{\delta^2 E}{\delta y^2}\right)^2 + \left(\frac{\delta^2 E}{\delta z^2}\right)^2} \qquad (11)$$

Where x, y and z are the considered atom's coordinates within the principal cluster.

Very recently, Spiridigliozzi et al. [13,15] proposed the application of the Cluster-plus-Glue Atom Model to include geometric and chemical features of constituting elements in the design of single-phase HEMs. In other words, the possibility for a hypothesized multicomponent system to be "written" at a given temperature into multiple principal clusters according to the Cluster-plus-Glue Atom Model rules, seems to be a necessary (yet not sufficient) condition for its entropy-driven stabilization at that temperature.

The Valence Combination Approach

Underlying chemical/geometric features similarly to the Cluster-plus-Glue Atom Model approach, Tang et al. [16] suggested a dual-stage approach, named Valence Combination approach, for more effectively exploring the landscape of High-Entropy Perovskite Oxides (HEPOs). Firstly, their method involves identifying the feasible independent valence combinations of the substituted cations, taking electrical stability into account. The material system is then segmented into various subsystems based on these identified valence combinations. In other words, materials that share the same substituted cation valence combinations are grouped under a single subsystem. Subsequently, appropriate cations for each subsystem are selected based on their geometric compatibility, including the use of factors such as Pauling's first rule and Goldschmidt's tolerance factor.

The rationale behind using valence combinations for classifying HEOs stems from the fact that the valence of substituted cations falls within specific integer ranges. Therefore, within any given system, the quantity of unique valence combinations—or subsystems—is limited, making the boundaries between these subsystems clearly definable.

Implementing this approach on the $Ba(5M_{0.2})O_3$ perovskite system, Tang et al. identified 12 unique subsystems, for which suitable substituted cations were determined. Of these, 10 compositions yielded single-phase cubic perovskites within a specific temperature range, whereas two exhibited (111) chemical ordering structures. The formation of a single-phase is predominantly influenced by the Goldschmidt tolerance factor, whereas the emergence of partially ordered structures is largely dictated by variations in cation valence. Their findings demonstrate that the Valence Combination approach holds promise for the effective design of high-entropy oxides and could be extended to additional oxide systems rather than HEPOs [17, 18].

Computational Theory and Methodologies

As already discussed, the assessment of phase stability and phase diagrams is facilitated through Gibbs free energy calculations. Various computational techniques are available for this purpose, including density functional theory (DFT), molecular dynamics simulations, and CALPHAD (Calculation of Phase Diagrams).

DFT-based methodologies are particularly effective for evaluating the energy of high-entropy materials, given their capacity to account for chemical complexity [19]. However, the limitation of DFT-based techniques lies in their computational demands, rendering them less effective for analyzing high-entropy materials with a complex mix of elements on a smaller scale.

Alternatively, molecular dynamics simulations offer an advantage in terms of computational speed and scalability, making them suitable for investigating materials composed of multiple elements [20, 21], even though they have a critical drawback as they do not account for interatomic potentials, making them less reliable for studying the complex chemistry of multi-component ceramics [22]. However, recent advancements in artificial intelligence (AI) and machine learning (ML) have paved the way for innovative computational models leveraging AI and ML algorithms to analyze DFT data for interatomic potentials [23]. Utilizing machine learning-based potentials allows for energy calculations to be performed in a timeframe comparable to molecular dynamics simulations while achieving an accuracy level akin to that of DFT-based methods.

In this context, several significant studies employing these computational techniques for predicting the stability and properties of high-entropy ceramics have been conducted till now [19-21, 24-26]. Importantly, some of these computational predictions have been corroborated through experimental data, underscoring the potential these computational methodologies hold for the future development of high-entropy materials.

Empirical Descriptors for High-Entropy Ceramics

Descriptors serve as analytical tools for outlining the general attributes or stability of a given system, guided by its basic characteristics and/or simple design parameters. The efficacy of these predictions heavily depends on the judicious selection of a fitting descriptor. Descriptors are generally sorted into broad categories: macro-scale, molecular-scale, and nano-scale. The choice of an optimal descriptor is informed by both specialized knowledge in physics and chemistry, and data-driven statistical analysis. Although descriptors have advanced significantly in multiple materials science domains, their application in high-entropy ceramics (HECs) is nascent.

One of the first descriptors introduced in the field of HECs has been derived by the Goldschmidt tolerance factor for scrutinizing the formation of single-phase perovskite oxides. Particularly, Jiang et al. [27] employed this factor as a descriptor when fabricating high-entropy perovskites, specifically those with high configurational entropy (≥ 1.5 R). Upon evaluating 12 high-entropy oxides with the general formula $(SrBa)(M)O_3$, they observed that a Goldschmidt tolerance factor close to one ($0.97 \leq t_G \leq 1.03$) is crucial for generating single-phases high-entropy perovskites, even though not being the sole requirement as also observed very recently by Spiridigliozzi et al. [15] for a high-entropy perovskite designed with the Cluster-plus-Glue Atom Model. Subsequently, Patel et al. [28] examined the impact of the Goldschmidt tolerance factor on the phase diagram and electronic properties of a specific perovskite thin film, observing that the tolerance factor t_G governs both the crystal architecture and the electronic behavior of perovskite-structured HECs.

Another simple descriptor was introduced by Spiridigliozzi et al. [29] for high-entropy oxides with either fluorite-like or bixbyite-like structures. Their research focused on 18 equiatomic five-element rare earth-based oxides, demonstrating that the standard deviation in the cationic radii of constituting elements served as a critical descriptor for their design. They concluded that a standard deviation in the distribution of cationic radii ($\delta > 0.095$) leads to a single-phase fluorite, whereas a value below 0.095 ($\delta < 0.095$) results in a single-phase bixbyite. Values of standard deviation very close to this threshold ($\delta \approx 0.095$) results in a prevalent fluorite-like system with a secondary bixbyite-like phase.

Similarly to Spiridigliozzi's descriptor based on standard deviation of the cationic radii (involving cations mismatch within the unit cell), Wright et al. [30] employed a descriptor known as the size disorder factor (δ_{size*}), calculated as the square root of the sum of the squares of lattice size differences at sites A and B in the pyrochlore structure, to forecast thermal conductivity of high-entropy oxides. Their results indicated that the size disorder factor is a more effective descriptor than ideal mixing entropy for predicting thermal conductivity in pyrochlore oxides. They observed that an increase in the size disorder factor led to a reduction in thermal conductivity, attributed to substantial lattice distortion.

Being another major focus of research in HECs, High-Entropy Carbides (HECs) have been intensively studied and many efforts have been made to identify descriptors that can predict their properties effectively. For example, Sarker et al. [31] employed what they termed as the "entropy-forming ability" (EFA) to evaluate the propensity of a material to form a single-phase high-entropy structure. The EFA analyzes the energy range of configurationally random calculations for unit cell formation. A high EFA value suggests a lower energy barrier for achieving randomness, thereby favoring higher entropy and disorder in the system.

Conversely, a lower EFA indicates a higher energy barrier to randomness, leading to more ordered phase formations. Through this approach, they studied 56 five-metal compositions using metals like Hf, Nb, Mo, Ta, Ti, V, W, and Zr. They found that higher EFA values are correlated with single-phase formations with reduced lattice distortion. They also highlighted that disorder enhances mechanical properties, making the EFA descriptor valuable for designing ultra-hard single-phase carbides with hardness levels of about 32-33 GPa. Subsequently, Harrington et al. [32] corroborated the use of entropy-forming ability, revealing that an EFA value greater than 45-50 eV/atom promotes the formation of an entropy-stabilized single phase, whereas lower values result in multiple phases.

Ning et al. [33] utilized the Vienna Ab-Initio Simulation Package (VASP) for Density Functional Theory (DFT) calculations and introduced lattice size difference, simply denoted as δ, as another descriptor for High-Entropy Carbides:

$$\delta = \sqrt{\frac{\sum_{i=0}^{n} x_i \left(1 - \dfrac{r_i}{\bar{r}}\right)^2}{n}} \tag{12}$$

Here, x_i, r_i and $\bar{r}$ represent the atomic fraction, the lattice constant of each component, and the average lattice constant for all components, respectively. Ning et al. observed that a smaller lattice size difference results in lower lattice distortion and a higher likelihood of the formation of a single-phase High-Entropy Carbide.

In one of the few endeavors to identify descriptors for high-entropy ceramics beyond High-Entropy Oxides and High-Entropy Carbides, Wen et al. [34] assessed the synthesis potential of a complex aluminosilicide composition using thermodynamic calculations and lattice size difference evaluations. Their results indicated that a single phase is probable in materials with a lattice size difference of 1.893%.

Finally, Liu et al. [35] conducted both computational and experimental studies on the likelihood of single-phase formation for a particular high-entropy boride composition. They reported that a lattice size difference of 2.763% is needed for single-phase formation, corroborating this assumption with experimental data on different single-phase High-Entropy Borides.

The utilization of various descriptors such as size mismatch, standard deviation of cationic radii, and lattice size differences in HECs underscores the crucial role that lattice geometry plays in determining not only the stability but also the properties of single-phase solutions. In other words, these descriptors can often be interpreted as metrics quantifying the degree of geometric compatibility or incompatibility among constituent elements, thus influencing single-phase formation, stability, and material properties of complex multicomponent systems. Similarly, descriptors based on disorder, like the size disorder factor or entropy-forming ability, focus on the extent of lattice distortion or the energetic favorability of disordered configurations. These descriptors, although different in their mathematical formulation, likewise reflect the system's geometric and structural considerations.

Definitely, the reliance on such descriptors signifies an overarching emphasis on the geometric configuration of the unit cell and the degree of disorder within a given multicomponent system. These are not isolated variables but interconnected factors that jointly contribute to our understanding and prediction of high-entropy ceramics. The convergence on geometry and disorder in these proposed descriptors suggests that these are fundamental aspects underpinning the design and synthesis of new high-entropy materials.

References

[1] Zhang, Y. & Zhou, Y.J. (2007, October). Solid solution formation criteria for high entropy alloys. *In:* Materials Science Forum (Vol. 561, pp. 1337–1339). Trans Tech Publications Ltd.

[2] Miedema, A.R., Boom, R. & De Boer, F.R. (1975). On the heat of formation of solid alloys. Journal of the Less Common Metals, 41(2), 283–298.

[3] Zhang, R.Z. & Reece, M.J. (2019). Review of high entropy ceramics: Design, synthesis, structure and properties. Journal of Materials Chemistry A, 7(39), 22148–22162.

[4] Poletti, M.G., Fiore, G., Szost, B.A. & Battezzati, L. (2015). Search for high entropy alloys in the X-NbTaTiZr systems (X = Al, Cr, V, Sn). Journal of Alloys and Compounds, 620, 283–288.

[5] Chen, R., Qin, G., Zheng, H., Wang, L., Su, Y., Chiu, Y., ... & Fu, H. (2018). Composition design of high entropy alloys using the valence electron concentration to balance strength and ductility. Acta Materialia, 144, 129–137.

[6] Takeuchi, A., Amiya, K., Wada, T., Yubuta, K. & Zhang, W. (2014). High-entropy alloys with a hexagonal close-packed structure designed by equi-atomic alloy strategy and binary phase diagrams. JOM, 66, 1984–1992.

[7] Li, R.X., Qiao, J.W., Liaw, P.K. & Zhang, Y. (2020). Preternatural hexagonal high-entropy alloys: A review. Acta Metallurgica Sinica (English Letters), 33, 1033–1045.

[8] Yang, X. & Zhang, Y. (2012). Prediction of high-entropy stabilized solid-solution in multi-component alloys. Materials Chemistry and Physics, 132(2–3), 233–238.

[9] Ye, Y.F., Wang, Q., Lu, J., Liu, C.T. & Yang, Y. (2015). Design of high entropy alloys: A single-parameter thermodynamic rule. Scripta Materialia, 104, 53–55.

[10] Dong, C., Wang, Q., Qiang, J.B., Wang, Y.M., Jiang, N., Han, G. & Xia, J.H. (2007). From clusters to phase diagrams: Composition rules of quasicrystals and bulk metallic glasses. Journal of Physics D: Applied Physics, 40(15), R273.

[11] Ma, Y., Dong, D., Wu, A. & Dong, C.(2018). Composition formulas of inorganic compounds in terms of cluster plus glue atom model. Inorganic Chemistry, 57(2), 710–717.

[12] Spiridigliozzi, L., Ferone, C., Cioffi, R. & Dell'Agli, G. (2023). Compositional design of single-phase rare-earth based high-entropy oxides (HEOs) by using the cluster-plus-glue atom model. Ceramics International, 49(5), 7662–7669.

[13] Miracle, D.B. (2004). A structural model for metallic glasses. Nature Materials, 3(10), 697–702.

[14] Du, J., Wen, B., Melnik, R. & Kawazoe, Y. (2014). Determining characteristic principal clusters in the "cluster-plus-glue-atom" model. Acta Materialia, 75, 113–121.

[15] Spiridigliozzi, L., Biesuz, M., Sglavo, V.M. & Dell'Agli, G. (2023). Design, synthesis and formation mechanism of a novel entropy-stabilized perovskite oxide derived from barium cerate/zirconate. Journal of the European Ceramic Society (in press).

[16] Tang, L., Li, Z., Chen, K., Li, C., Zhang, X. & An, L. (2021). High-entropy oxides based on valence combinations: Design and practice. Journal of the American Ceramic Society, 104(5), 1953–1958.

[17] Ma, J., Chen, K., Li, C., Zhang, X. & An, L. (2021). High-entropy stoichiometric perovskite oxides based on valence combinations. Ceramics International, 47(17), 24348–24352.

[18] Jia, H., Li, C., Chen, G., Li, H., Li, S., An, L. & Chen, K. (2022). Design and synthesis of high-entropy pyrochlore ceramics based on valence combination. Journal of the European Ceramic Society, 42(13), 5973–5983.

[19] Pitike, K.C., Kc, S., Eisenbach, M., Bridges, C.A. & Cooper, V.R. (2020). Predicting the phase stability of multicomponent high-entropy compounds. Chemistry of Materials, 32(17), 7507–7515.

[20] Lim, M., Rak, Z., Braun, J.L., Rost, C.M., Kotsonis, G.N., Hopkins, P.E., ... & Brenner, D.W. (2019). Influence of mass and charge disorder on the phonon thermal conductivity of entropy stabilized oxides determined by molecular dynamics simulations. Journal of Applied Physics, 125(5).

[21] Jiang, B., Bridges, C.A., Unocic, R.R., Pitike, K.C., Cooper, V.R., Zhang, Y., ... & Page, K. (2020). Probing the local site disorder and distortion in pyrochlore high-entropy oxides. Journal of the American Chemical Society, 143(11), 4193–4204.

[22] Akrami, S., Edalati, P., Fuji, M. & Edalati, K. (2021). High-entropy ceramics: Review of principles, production and applications. Materials Science and Engineering: R: Reports, 146, 100644.

[23] Dai, F.Z., Wen, B., Sun, Y., Xiang, H. & Zhou, Y. (2020). Theoretical prediction on thermal and mechanical properties of high entropy $(Zr_{0.2}Hf_{0.2}Ti_{0.2}Nb_{0.2}Ta_{0.2})$ C by deep learning potential. Journal of Materials Science & Technology, 43, 168–174.

[24] Svane, K.L. & Rossmeisl, J. (2022). Theoretical optimization of compositions of high-entropy oxides for the oxygen evolution reaction. Angewandte Chemie, 134(19), e202201146.

[25] Sharma, Y., Lee, M.C., Pitike, K.C., Mishra, K.K., Zheng, Q., Gao, X., ... & Ward, T.Z. (2022). High entropy oxide relaxor ferroelectrics. ACS Applied Materials & Interfaces, 14(9), 11962–11970.

[26] Vyatskikh, A.L., MacDonald, B.E., Dupuy, A.D., Lavernia, E.J., Schoenung, J.M. & Hahn, H. (2021). High entropy silicides: CALPHAD-guided prediction and thin film fabrication. Scripta Materialia, 201, 113914.

[27] Jiang, S., Hu, T., Gild, J., Zhou, N., Nie, J., Qin, M., ... & Luo, J. (2018). A new class of high-entropy perovskite oxides. Scripta Materialia, 142, 116–120.

[28] Patel, R.K., Ojha, S.K., Kumar, S., Saha, A., Mandal, P., Freeland, J.W. & Middey, S. (2020). Epitaxial stabilization of ultra thin films of high entropy perovskite. Applied Physics Letters, 116(7).

[29] Spiridigliozzi, L., Ferone, C., Cioffi, R. & Dell'Agli, G. (2021). A simple and effective predictor to design novel fluorite-structured High Entropy Oxides (HEOs). Acta Materialia, 202, 181–189.

[30] Wright, A.J., Wang, Q., Ko, S.T., Chung, K.M., Chen, R. & Luo, J. (2020). Size disorder as a descriptor for predicting reduced thermal conductivity in medium- and high-entropy pyrochlore oxides. Scripta Materialia, 181, 76–81.

[31] Sarker, P., Harrington, T., Toher, C., Oses, C., Samiee, M., Maria, J.P., ... & Curtarolo, S. (2018). High-entropy high-hardness metal carbides discovered by entropy descriptors. Nature Communications, 9(1), 4980.

[32] Harrington, T.J., Gild, J., Sarker, P., Toher, C., Rost, C.M., Dippo, O.F., ... & Vecchio, K.S. (2019). Phase stability and mechanical properties of novel high entropy transition metal carbides. Acta Materialia, 166, 271–280.

[33] Ning, S., Wen, T., Ye, B. & Chu, Y. (2020). Low-temperature molten salt synthesis of high-entropy carbide nanopowders. Journal of the American Ceramic Society, 103(3), 2244–2251.

[34] Wen, T., Liu, H., Ye, B., Liu, D. & Chu, Y. (2020). High-entropy alumino-silicides: A novel class of high-entropy ceramics. Science China Materials, 63(2), 300–306.

[35] Liu, D., Liu, H., Ning, S. & Chu, Y. (2020). Chrysanthemum-like high-entropy diboride nanoflowers: A new class of high-entropy nanomaterials. Journal of Advanced Ceramics, 9, 339–348.

Noticeable High-Entropy Ceramics' Thermal Properties

Thermal Properties of Materials

Thermal properties pertain to how a substance behaves when exposed to heat. When heat energy is introduced to a solid, the result is a rise in its temperature and a consequent expansion in its size. If there are temperature differences within the specimen, the energy may move to its colder areas, and eventually, the specimen might experience a fusion. In many practical applications of solids, properties such as heat capacity, thermal expansion, and thermal conductivity are pivotal.

Heat Capacity

Upon heating, a solid sees a temperature upsurge, indicating its absorption of a certain amount of energy. The capacity of a material to *draw in heat* from its surroundings is depicted by its heat capacity. Heat capacity quantifies the energy amount necessary to achieve a single-degree temperature increment. Mathematically, heat capacity (C) can be represented as:

$$C = \frac{dQ}{dT} \tag{1}$$

In Equation (1), dQ represents the energy needed to bring a dT temperature variation. Commonly, heat capacity is denoted per mole of substance (J/mol·K or cal/mol·K).

Specific heat (often represented by a lowercase c) refers to the heat capacity on a per mass basis, and its units can vary according to different disciplines (i.e. J/kg·K, cal/g·K, Btu/lbm·°F).

Thermal Expansion

Solid substances mostly tend to enlarge when heated and shrink when cooled. The alteration in length due to temperature change in a solid can be described by:

$$\frac{l_f - l_0}{l_0} = \alpha_l \, (T_f - T_0)$$ (2)

or

$$\frac{\Delta l}{l_0} = \alpha_l \Delta T$$ (3)

In these equations, l_0 and l_f stand for the initial and final lengths when the temperature shifts from T_0 to T_f. The term α_l represents the linear coefficient of thermal expansion and indicates the degree to which a substance enlarges upon being heated. It possesses units of inverse temperature, specifically $[(°C)^{-1}$ or $(°F)^{-1}]$.

However, temperature variations influence not just the length but all spatial dimensions of an object, leading to an overall change in its volume. The computation for volume alterations due to temperature is:

$$\frac{\Delta V}{V_0} = \alpha_v \Delta T$$ (4)

In Equation (4), ΔV represents the volume change, while V_0 stands for the initial volume. The term α_v is the volume coefficient of thermal expansion. It's worth noting that for many substances, α_v exhibits an anisotropic behavior, meaning its value varies based on the crystallographic direction in which it's assessed.

In cases where a material's thermal expansion is isotropic (i.e. uniform in all directions), the relation is such that α_v is roughly equal to three times α_l.

Thermal Conductivity

Thermal conduction is the process through which heat is transferred from regions with higher temperatures to those with lower temperatures within a material. The property determining a material's aptitude to conduct heat is defined as its thermal conductivity. This property can be measured using the following equation:

$$q = -k\frac{dT}{dx}$$ (5)

Here, q stands for the heat flux, implying the heat flow on a per unit area basis, considering the area perpendicular to the direction of flow; k stands for the thermal conductivity, while dT/dx indicates the temperature gradient traversing the medium of conduction.

The measure units used for q and k are W/m² (or Btu/ft²·h) and W/m·K (or Btu/ft·h·°F), respectively. The validity of Equation (5) is limited to conditions with a steady-state heat flow, i.e. in scenarios wherein the heat flux remains unchanged over time.

The negative sign in Equation (5) indicates that heat transport always occurs from warmer regions towards cooler ones, aligning with the temperature gradient. A point to note is that Equation (5) bears resemblance to Fick's first law [1] used in steady-state diffusion contexts:

$$J = -D\frac{dC}{dx} \tag{6}$$

where k parallels the diffusion coefficient D, and the temperature gradient can be compared to the concentration gradient, represented as dC/dx.

Heat Conduction Mechanisms

Within solid materials, heat transmission occurs chiefly through lattice vibration waves, known as phonons, and through free electrons. Each of these mechanisms possesses its own associated thermal conductivity, and the composite conductivity can be summarized as:

$$k = k_l + k_e \tag{7}$$

In Equation (7), k_l and k_e represent the thermal conductivities resulting from lattice vibrations and electrons, respectively. Generally, one of these two factors holds dominance over the other one.

The thermal energy connected with phonons or lattice waves moves in their direction of propagation. The k_l component emerges from a predominant shift of phonons from warmer to cooler zones within a material subjected to a temperature gradient.

Free electrons, which can move about, play a role in electronic thermal conduction. When these electrons are in a material's warmer section, they accumulate kinetic energy. Their subsequent migration to cooler parts leads to a transfer of some of this energy to atoms (in the form of vibrational energy) due to collisions with phonons or other lattice irregularities. As the concentration of free electrons rises, the contribution of k_e to overall thermal conductivity amplifies, given that more electrons are present to aid in the heat transfer mechanism.

Thermal Stresses

Thermal stresses refer to stresses developed within a material due to temperature fluctuations. It is pivotal to understand the genesis and characteristics of these kind of stresses as they have the potential to cause fractures or unintended plastic deformations within materials.

To delve into thermal stresses, a uniform, isotropic solid rod can be considered as an exemplary specimen. When such rod undergoes uniform heating or cooling without temperature differentials, it would expand or contract freely without any stress under normal circumstances. But, if there are rigid supports at the rod's ends that restrict its axial movement, then thermal stresses come into play. To quantify the stress (denoted by σ) brought about by a temperature shift from T_0 to T_f, it is possible to refer to the following equation:

$$\sigma = E\alpha_l \, (T_0 - T_f) \tag{8}$$

In Equation (8), E stands for the elasticity modulus, and α_l is the linear coefficient of thermal expansion.

In scenarios where the exemplary rod is subjected to heating ($T_f > T_0$), a compressive stress arises ($\sigma < 0$) because the natural expansion of the rod is hindered. Conversely, when the rod is cooled ($T_f < T_0$), a tensile stress is engendered ($\sigma > 0$).

It is noteworthy to mention that the stress magnitude defined in Equation (8) matches the stress necessary to elastically revert the rod to its original length. In other words, this is after permitting the rod to undergo free expansion or contraction due to the $T_0 - T_f$ temperature variation.

Thermal Properties of High-entropy Ceramics

The high levels of lattice distortion and notable mass variations in high-entropy oxides results in their thermal conductivity being comparatively lower than those of non-high-entropy analogs. As an example, the combined thermal conductivity and the lattice thermal conductivity for undiluted $SrTiO_3$ were fairly elevated, lying between 9.79 and 5.56 W/(m·K) for temperatures ranging from 323 to 873 K [2]. Upon a 10% La and 10% Nb doping [3], the thermal conductivity of $(Sr_{0.9}La_{0.1})(Ti_{0.9}Nb_{0.1})O_3$ was reduced. Yet, its collective thermal conductivity and lattice thermal conductivity remained above 2.9 and 2.3 W/(m·K) at 873 K, respectively. Conversely, for the high-entropy $(Ca_{0.2}Sr_{0.2}Ba_{0.2}Pb_{0.2}La_{0.2})TiO_3$ perovskite oxide, thermal conductivity was between 1.7 and 1.0 W/(m·K), which was significantly lower than the reference materials [4].

Typically, thermal conductivity is strongly linked to both mass and strain variances (characterized by disorder scattering factors). Specifically, an increase in mass and strain fluctuations leads to a decline in both the phonon relaxation time and the thermal conductivity of a material [5].

As high-entropy ceramics own significant disorder, pronounced lattice distortion and large mass disparities at crystallographic sites, they are potential candidates for materials with reduced thermal conductivity.

Oxygen vacancies also have a significant influence on thermal conductivity: when such vacancies arose, they functioned as defects, scattering short-wavelength phonons. Concurrently, they contribute to the entropy at the oxygen site. Furthermore, only minor variances in thermal conductivity under different processing conditions of the same high-entropy material has been observed [4]. This could be attributed to a consistent contribution of oxygen vacancies to phonon scattering during the annealing of samples.

However, even though many researchers observed that high-entropy ceramics (HECs) exhibit reduced thermal conductivities [6-9], few research studies have delved into the specific underlying mechanisms or major influencing factors. Braun et al. [10] suggested that this reduction in six-component rocksalt oxides could be attributed to disorder in force constants resulting from charge fluctuations. A similar assertion by Yang et al. [11] was made for Ln_3NbO_7, emphasizing force-constant disorder due to size disorder and charge variation. It was observed in both these studies that there was a high ratio of Young's modulus to thermal conductivity (E/k), signaling a considerable phonon scattering rate.

A recent proposition suggested extending the definition of HECs to cover compositionally-complex ceramics (CCCs) or multi-principal cation ceramics (MPCCs) too [12]. This expansion is due to the observation that medium-entropy ceramics might sometimes outperform their high-entropy counterparts in terms of having a lower thermal conductivity and a higher E/k.

To probe into the critical determinant governing the diminished thermal conductivity in HECs and by extension, CCCs, an examination was conducted by Wright et al. [13] on 22 single-phase pyrochlore oxides. This comprised 18 compositionally-complex (medium- and high-entropy) pyrochlore oxides (CCPOs) and four "low-entropy" benchmark variants. According to definitions in [12], "high entropy" refers to ceramics with five or more main cations, while "medium entropy" alludes to those with three or four principal cations or non-equimolar compositions.

Interestingly, thermal conductivity seems to be more closely associated with size disorder rather than entropy itself.

Pyrochlores with the $A_2B_2O_7$ structure can be viewed as structured fluorites with two distinct cation layers (see the dedicated chapter on fluorites and fluorite-structured entropy oxides), positioning them as potential contenders to yttria-stabilized zirconia (YSZ) for thermal barrier coatings (TBCs) [14]. Pyrochlores in their cubic form possess fairly high Young's moduli (E ~250 GPa) [15, 16] and exhibit low heat conductivities (k ~2 W m^{-1} K^{-1}) [17, 18]. Efforts to further decrease thermal conductivity have looked at adding another rare earth element to the A-site [19-22] or B-site [23]. Recent works have highlighted the creation of single-phase high-entropy zirconate-based pyrochlores with reduced thermal conductivities, even considering potential porosity and structural factors [24, 25].

In their recent research, Wright et al. [13] produced over 20 dense CCPOs, modifying mass and size disorder across both A and B sites, also introducing a descriptor to predict the thermal conductivity reduction employing the mass disorder (g) and size disorder (δ) parameters traditionally used in thermal transportation and high-entropy alloys (HEAs) studies [26, 27].

From Wright et al. data, it is evident that most CCPOs had reduced thermal conductivities compared to benchmark specimens, noting up to 35% drops. However, some exceptions have been observed, such as $La_2(Hf_{0.5}Zr_{0.5})_2O_7$ which had a slightly higher conductivity than its non-high-entropy analogous. Despite such isolated points, a general trend of decreased thermal conductivity, with moduli retention, has been observed. This simultaneous reduction in conductivity and maintenance of modulus is a noteworthy aspect for HECs. The authors of the study [13] found relationships between the three adopted descriptors (δ* size, g*, ρ) and analyzed the obtained k and E values.

Linear relationships were determined, and correlations were assessed through the Pearson correlation coefficient (PCC) and r^2.

Figure 1 indicates a strong correlation between thermal conductivity (k) and the size disorder, denoted as δ*size.

A noticeable negative relationship exists between δ*size and k, with a Pearson Correlation Coefficient (PCC) of –0.86 and a coefficient of determination (r^2) of 0.73.

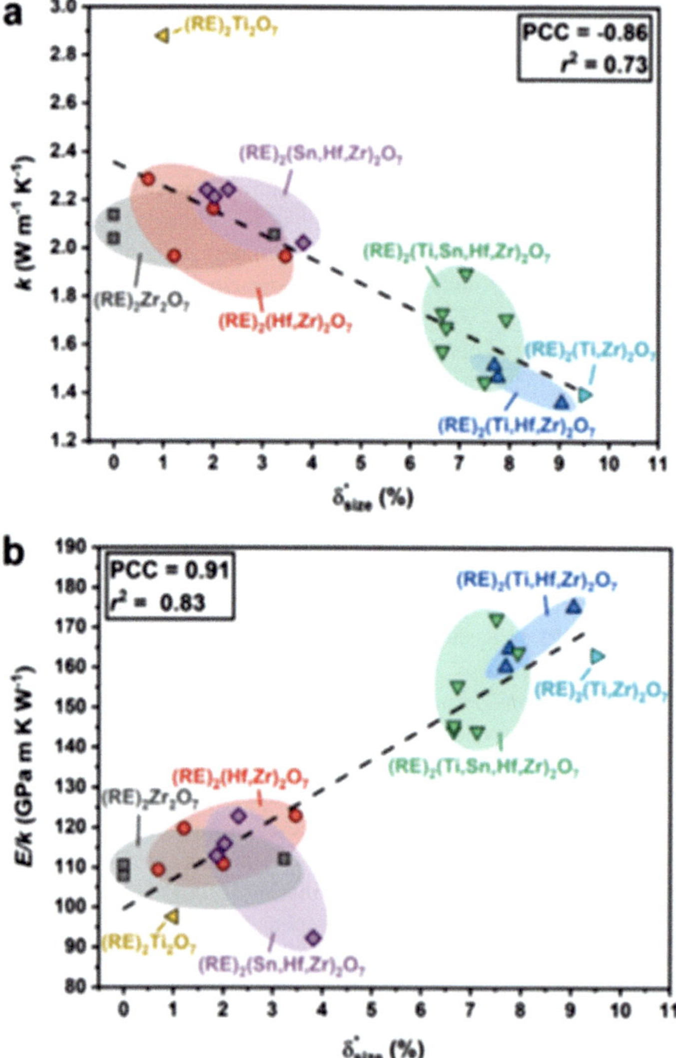

Fig. 1: Correlation of (a) thermal conductivity (k) and (b) the *E/k* ratios of all 22 single-phase pyrochlores made in this study with the size disorder parameter, δ*size. Figure taken from [13].

The equation representing their linear relationship is given by:

$$k = 2.36 - 10\,\mathrm{d} * \text{size} \,(\mathrm{Wm^{-1}K^{-1}}) \tag{9}$$

Additionally, k has a closer relation to $\delta*$size, than to individual size disorders δA and δB for each sublattice or the total disorder assuming random mixing. Additional analyses have revealed that the B-site size disorder (δB) surpasses the A-site's (δA), making it more effective at reducing k. This observation aligns with previous models referenced [28, 29].

Anyway, a perfect linear relationship between k and any single descriptor used by Wright et al. has not been demonstrated due to various influencing factors.

Firstly, there is an insignificant correlation between $\delta*$size and E (PCC = 0.13, r^2 = 0.02), suggesting that introducing size disorder in CCPOs maintains a high E modulus. Secondly, an evident positive relationship exists between the E/k ratio and $\delta*$size (PCC = 0.91, r^2 = 0.83) with the following defining equation:

$$E/k = 99.6 + 750\,\delta * \text{size} \,(\mathrm{GPam\ KW^{-1}}) \tag{10}$$

When examining the mass disorder g*, it seems to be less influential than $\delta*$size, especially in the context of Ti-containing compounds. Correlations with density ρ seem lacking.

Interestingly, despite expected links between size disorder and mixing entropy, particularly on the B site, mixing configurational entropies generally show weaker associations with thermal conductivities.

Definitely, size disorder seems to offer a more effective approach to designing low-k CCCs than mixing configurational entropy, emphasizing lattice distortion's importance over "high entropy" in decreasing both HECs and CCCs' thermal conductivity.

Moreover, all HECs and CCCs compositions incorporating Ti had a $\delta*$size exceeding 6% and a lower k, leading to a substantially increased E/k ratio. Titanium's properties, as both a smaller and lighter element, not only augment size disorder but also diminish k, acting as a "rattler" on the B-site. This further establishes that ionic radius can be more influential in modulating k than atomic mass, resulting in high E/k ratios for Ti-enriched compositions, with $(\mathrm{Sm_{0.33}}$ $\mathrm{Eu_{0.33}Gd_{0.33})_2(Ti_{0.25}Sn_{0.25}Hf_{0.25}Zr_{0.25})_2O_7}$ being a peculiar example.

To delve deeper into the influence of Ti, it has been observed [13] that k is at its lowest when [Ti] = 0.25, whereas the modulus E rises with increasing Ti content, especially when Ti occupies around half of the B-site. The "rattler" effect, previously observed in pyrochlores, was further extended in analyzing Yb's potential as a "rattler" on the A-site. Specifically, introducing Yb in the A-site resulted in a roughly 13% drop in k with a minor decline in E, culminating in a high E/k ratio [13]. Thus, the reduced k in the aforementioned HECs might be influenced also by microstructural factors.

To summarize, size disorder in HECs and CCCs can lead to both severely deformed sublattices and enlarged cavities suitable for the "rattler" effect. Both these factors can significantly impede heat conduction. Additionally, unique to the pyrochlore structure, the oxygen ion at the 8a-site can shift to the oxygen vacancy position, potentially leading to a stressed anion sublattice. In this context, the size disorder can be enhanced when combining the largest and smallest elements in a four-component mix for each sublattice.

Concluding Remarks

High-entropy oxides, particularly those exhibiting the pyrochlore structure, have sparked significant interest in the materials science community, primarily due to their unique thermal properties.

The most dominant factor affecting thermal conductivity (k) appears to be the size disorder (δ*size). This disorder can lead to both severely deformed sublattices and enlarged cavities suitable for the rattler effect, which can notably impede heat conduction. The size disorder can be enhanced when combining the largest and smallest elements in a four-component mix on each sublattice.

When size disorder is introduced into CCCs, a positive correlation emerges between the E/k ratio and δ*size, highlighting an intricate relationship between elastic modulus and thermal conductivity as mediated by size disorder.

Conversely, the mass disorder appears to be a less potent descriptor for thermal properties compared to size disorder. Notably, density (ρ) exhibits weak correlations with both k and the E/k ratio. Mixing configurational entropies have also been shown to exhibit weaker correlations with thermal conductivities.

The pyrochlore structure brings about specific phenomena. For instance, the oxygen ion at the 8a-site in the structure can shift to an oxygen vacancy position, potentially leading to a stressed anion sublattice. Such structural modifications could further influence thermal properties. Moreover, specific elements can exert a pronounced influence on high-entropy pyrochlores. Titanium (Ti), for example, seems to bring unique properties over CCCs. Very likely, titanium's smaller and lighter nature, in comparison to other metal elements, accentuates the size disorder, acting as a "rattler" on the B-site, which suppresses k.

Table 1 shows most of the reported HECs and CCCs exhibiting noticeable thermal conductivities.

It is crucial to note discrepancies in thermal properties' measurements across different studies, as demonstrated by the differences in reported thermal conductivities for similar compositions.

This emphasizes the importance of ensuring sample purity, eliminating microstructure effects, and maintaining high relative densities to achieve accurate and reliable measurements. Anyway, to the state-of-the-art, many pyrochlore-structured, fluorite-structured and/or perovskite-structured HECs and CCCs hold huge promise for use in near-future technological applications related to thermal properties, such as advanced thermal barrier coatings for extreme-high-temperature environments.

Table 1: Summary of the most promising HECs/CCCs in terms of their thermal properties

Composition	k [W/mK]	Reference
$(La_{0.2}Ce_{0.2}Nd_{0.2}Sm_{0.2}Eu_{0.2})_2Zr_2O_7$	2.06	[13]
$(La_{0.14}Ce_{0.14}Pr_{0.14}Nd_{0.14}Sm_{0.14}Eu_{0.14}Gd_{0.14})_2(Hf_{0.5}Zr_{0.5})_2O_7$	1.97	[13]
$(La_{0.14}Ce_{0.14}Pr_{0.14}Nd_{0.14}Sm_{0.14}Eu_{0.14}Gd_{0.14})_2(Sn_{0.33}Hf_{0.33}Zr_{0.33})_2O_7$	2.02	[13]
$(Sm_{0.25}Eu_{0.25}Gd_{0.25}Yb_{0.25})_2(Ti_{0.25}Sn_{0.25}Hf_{0.25}Zr_{0.25})_2O_7$	1.45	[13]
$(Sm_{0.33}Eu_{0.33}Gd_{0.33})_2(Ti_{0.25}Sn_{0.25}Hf_{0.25}Zr_{0.25})_2O_7$	1.67	[13]
$Sm_2(Sn_{0.25}Ti_{0.25}Hf_{0.25}Zr_{0.25})_2O_7$	1.73	[13]
$Gd_2(Sn_{0.25}Ti_{0.25}Hf_{0.25}Zr_{0.25})_2O_7$	1.58	[13]
$(Sm_{0.5}Gd_{0.5})_2(Ti_{0.33}Hf_{0.33}Zr_{0.33})_2O_7$	1.47	[13]
$(Ce_{0.2}Zr_{0.2}Hf_{0.2}Sn_{0.2}Ti_{0.2})O_2$	1.28	[30]
$(Hf_{0.25}Zr_{0.25}Ce_{0.25})(Y_{0.125}Ca_{0.125})O_2$	1.10	[9]
$Mg_{0.2}Ni_{0.2}Cu_{0.2}Co_{0.2}Zn_{0.2}O$	2.95	[7]
$Mg_{0.167}Ni_{0.167}Cu_{0.167}Co_{0.167}Zn_{0.167}Sc_{0.167}O$	1.68	[7]
$Mg_{0.167}Ni_{0.167}Cu_{0.167}Co_{0.167}Zn_{0.167}Sb_{0.167}O$	1.41	[7]
$Mg_{0.167}Ni_{0.167}Cu_{0.167}Co_{0.167}Zn_{0.167}Sn_{0.167}O$	1.44	[7]
$Mg_{0.167}Ni_{0.167}Cu_{0.167}Co_{0.167}Zn_{0.167}Cr_{0.167}O$	1.64	[7]
$Mg_{0.167}Ni_{0.167}Cu_{0.167}Co_{0.167}Zn_{0.167}Ge_{0.167}O$	1.60	[7]
$(Dy_{0.2}Ho_{0.2}Er_{0.2}Y_{0.2}Yb_{0.2})_3NbO_7$	0.72	[31]
$(Hf_{0.2}Zr_{0.2}Ce_{0.2}Y_{0.2}Yb_{0.2})O_2$	2.23	[32]
$(Hf_{0.314}Zr_{0.314}Ce_{0.314}Y_{0.029}Ca_{0.029})O_2$	1.65	[32]
$(Hf_{0.284}Zr_{0.284}Ce_{0.284}Y_{0.074}Ca_{0.074})O_2$	1.54	[32]
$(Hf_{0.314}Zr_{0.314}Ce_{0.314}Y_{0.029}Gd_{0.029})O_2$	1.74	[32]
$(Hf_{0.284}Zr_{0.284}Ce_{0.284}Y_{0.074}Gd_{0.074})O_2$	1.70	[32]
$(Y_{0.25}Yb_{0.25}Er_{0.25}Lu_{0.25})_2(Zr_{0.5}Hf_{0.5})_2O_7$	1.40	[33]
$(Ca_{0.2}Sr_{0.2}Ba_{0.2}Pb_{0.2}La_{0.2})TiO_3$	1.17	[34]
$Sr(Ti_{0.2}Fe_{0.2}Mo_{0.2}Nb_{0.2}Cr_{0.2})O_3$	0.7	[35]

References

[1] Callister Jr, W.D. & Rethwisch, D.G. (2020). Fundamentals of Materials Science and Engineering: An Integrated Approach. John Wiley & Sons.

[2]   Popuri, S.R., Scott, A.J.M., Downie, R.A., Hall, M.A., Suard, E., Decourt, R., ... & Bos, J.W. (2014). Glass-like thermal conductivity in $SrTiO_3$ thermoelectrics induced by A-site vacancies. RSC Advances, 4(64), 33720-33723.

[3]   Wang, J., Zhang, B.Y., Kang, H.J., Li, Y., Yaer, X., Li, J.F., ... & Zhao, L.D. (2017). Record high thermoelectric performance in bulk $SrTiO_3$ via nano-scale modulation doping. Nano Energy, 35, 387-395.

[4]   Zheng, Y., Zou, M., Zhang, W., Yi, D., Lan, J., Nan, C.W. & Lin, Y.H. (2021). Electrical and thermal transport behaviours of high-entropy perovskite thermoelectric oxides. Journal of Advanced Ceramics, 10, 377-384.

[5]   Ren, G.K., Lan, J.L., Ventura, K.J., Tan, X., Lin, Y.H. & Nan, C.W. (2016). Contribution of point defects and nano-grains to thermal transport behaviours of oxide-based thermoelectrics. NPJ Computational Materials, 2(1), 1-9.

[6]   Gild, J., Braun, J., Kaufmann, K., Marin, E., Harrington, T., Hopkins, P., ... & Luo, J. (2019). A high-entropy silicide: $(Mo_{0.2}Nb_{0.2}Ta_{0.2}Ti_{0.2}W_{0.2})Si_2$. Journal of Materiomics, 5(3), 337-343.

[7]   Braun, J.L., Rost, C.M., Lim, M., Giri, A., Olson, D.H., Kotsonis, G.N., ... & Hopkins, P.E. (2018). Charge-induced disorder controls the thermal conductivity of entropy-stabilized oxides. Advanced Materials, 30(51), 1805004.

[8]   Liu, R., Chen, H., Zhao, K., Qin, Y., Jiang, B., Zhang, T., ... & Chen, L. (2017). Entropy as a gene-like performance indicator promoting thermoelectric materials. Advanced Materials, 29(38), 1702712.

[9]   Gild, J., Samiee, M., Braun, J.L., Harrington, T., Vega, H., Hopkins, P.E., ... & Luo, J. (2018). High-entropy fluorite oxides. Journal of the European Ceramic Society, 38(10), 3578-3584.

[10]   Braun, J.L., Rost, C.M., Lim, M., Giri, A., Olson, D.H., Kotsonis, G.N., ... & Hopkins, P.E. (2018). Charge-induced disorder controls the thermal conductivity of entropy-stabilized oxides. Advanced Materials, 30(51), 1805004.

[11]   Yang, J., Qian, X., Pan, W., Yang, R., Li, Z., Han, Y., ... & Wan, C. (2019). Diffused lattice vibration and ultralow thermal conductivity in the binary Ln–Nb–O oxide system. Advanced Materials, 31(24), 1808222.

[12]   Wright, A.J., Wang, Q., Huang, C., Nieto, A., Chen, R. & Luo, J. (2020). From high-entropy ceramics to compositionally-complex ceramics: A case study of fluorite oxides. Journal of the European Ceramic Society, 40(5), 2120-2129.

[13]   Wright, A.J., Wang, Q., Ko, S.T., Chung, K.M., Chen, R. & Luo, J. (2020). Size disorder as a descriptor for predicting reduced thermal conductivity in medium- and high-entropy pyrochlore oxides. Scripta Materialia, 181, 76-81.

[14]   Wu, J., Wei, X., Padture, N.P., Klemens, P.G., Gell, M., García, E., ... & Osendi, M.I. (2002). Low-thermal-conductivity rare-earth zirconates for potential thermal-barrier-coating applications. Journal of the American Ceramic Society, 85(12), 3031-3035.

[15]   Pruneda, J.M. & Artacho, E. (2005). First-principles study of structural, elastic, and bonding properties of pyrochlores. Physical Review B, 72(8), 085107.

[16]   Yang, J., Shahid, M., Zhao, M., Feng, J., Wan, C. & Pan, W. (2016). Physical properties of $La_2B_2O_7$ (BZr, Sn, Hf and Ge) pyrochlore: First-principles calculations. Journal of Alloys and Compounds, 663, 834-841.

[17]   Clarke, D.R. & Phillpot, S.R. (2005). Thermal barrier coating materials. Materials Today, 8(6), 22-29.

[18] Vassen, R., Cao, X., Tietz, F., Basu, D. & Stöver, D. (2000). Zirconates as new materials for thermal barrier coatings. Journal of the American Ceramic Society, 83(8), 2023-2028.

[19] Liu, Z.G., Ouyang, J.H. & Zhou, Y. (2009). Structural evolution and thermophysical properties of $(Sm_xGd1{-}x)_2Zr_2O_7$ ($0{\leq}x{\leq}1.0$) ceramics. Journal of Alloys and Compounds, 472(1-2), 319-324.

[20] Zhang, H., Kun, S.U.N., Qiang, X., Fuchi, W.A.N.G. & Ling, L.I.U. (2009). Thermal conductivity of $(Sm1{-}xLax)_2Zr_2O_7$ (x = 0, 0.25, 0.5, 0.75 and 1) oxides for advanced thermal barrier coatings. Journal of Rare Earths, 27(2), 222-226.

[21] Pan, W., Wan, C.L., Xu, Q., Wang, J.D. & Qu, Z.X. (2007). Thermal diffusivity of samarium–gadolinium zirconate solid solutions. Thermochimica Acta, 455(1-2), 16-20.

[22] Wang, Y., Yang, F. & Xiao, P. (2013). Rattlers or oxygen vacancies: Determinant of high temperature plateau thermal conductivity in doped pyrochlores. Applied Physics Letters, 102(14).

[23] Wan, C., Qu, Z., Du, A. & Pan, W. (2009). Influence of B site substituent Ti on the structure and thermophysical properties of $A_2B_2O_7$-type pyrochlore $Gd_2Zr_2O_7$. Acta Materialia, 57(16), 4782-4789.

[24] Guo, Y., Feng, S., Yang, Y., Zheng, R., Zhang, Y., Fu, J., ... & Li, J. (2022). High-entropy titanate pyrochlore as newly low-thermal conductivity ceramics. Journal of the European Ceramic Society, 42(14), 6614-6623.

[25] Ma, W., Luo, Y., Ma, Z., Li, C., Qian, Y., Li, W., ... & Yang, J. (2023). $La_2Zr_2O_7$ based high entropy ceramic with a low thermal conductivity and enhanced CMAS corrosion resistance. Ceramics International, 49(18), 29729-29735.

[26] Zhang, Y., Zhou, Y., Lin, J., Chen, G. & Liaw, P. (2008), Solid-solution phase formation rules for multi-component alloys. Advanced Engineering Materials, 10: 534-538.

[27] Wu, C.-S., Tsai, P.-H., Kuo, C.-M. & Tsai, C.-W. (2018). Effect of atomic size difference on the microstructure and mechanical properties of high-entropy alloys. Entropy, 20(12), 967.

[28] Schelling, P.K., Phillpot, S.R. & Grimes, R.W. (2004). Optimum pyrochlore compositions for low thermal conductivity. Philosophical Magazine Letters, 84(2), 127-137.

[29] Minervini, L., Grimes, R.W. and Sickafus, K.E. (2000).Disorder in pyrochlore oxides. Journal of the American Ceramic Society, 83, 1873-1878.

[30] Chen, K., Pei, X., Tang, L., Cheng, H., Li, Z., Li, C. et al. (2018). A five-component entropy stabilized fluorite oxide. Journal of the European Ceramic Society, 38, 4161-4164.

[31] Zhu, J., Meng, X., Xu, J., Zhang, P., Lou, Z., Reece, M.J. & Gao, F. (2021). Ultra-low thermal conductivity and enhanced mechanical properties of high-entropy rare earth niobates (RE_3NbO_7, RE = Dy, Y, Ho, Er, Yb). Journal of the European Ceramic Society, 41, 1052-1057.

[32] Wright, A.J., Wang, Q., Huang, C., Nieto, A., Chen, R. & Luo, J. (2020). From high-entropy ceramics to compositionally-complex ceramics: A case study of fluorite oxides. Journal of the European Ceramic Society, 40, 2120-2129.

[33] Zhao, Z., Chen, H., Xiang, H., Dai, F.-Z., Wang, X., Xu, W. et al. (2020). $(Y_{0.25}Yb_{0.25}Er_{0.25}Lu_{0.25})_2(Zr_{0.5}Hf_{0.5})_2O_7$: A defective fluorite structured high entropy ceramic

with low thermal conductivity and close thermal expansion coefficient to Al_2O. Journal of Materials Science & Technology, 39, 167-172.

[34] Zheng, Y., Zou, M., Zhang, W., Yi, D., Lan, J., Nan, C.W. & Lin, Y.H. (2021). Electrical and thermal transport behaviours of high-entropy perovskite thermoelectric oxides. Journal of Advanced Ceramics, 10, 377-384.

[35] Banerjee, R., Chatterjee, S., Ranjan, M., Bhattacharya, T., Mukherjee, S., Jana, S.S., ... & Maiti, T. (2020). High-entropy perovskites: An emergent class of oxide thermoelectrics with ultralow thermal conductivity. ACS Sustainable Chemistry & Engineering, 8(46), 17022-17032.

Noticeable High-Entropy Ceramics' Electrical Properties

Electrical Properties of Materials

One of the most important electrical property of a solid material is its electrical conductance, i.e. the facility with which a given material enables the traversal of an electrical current. The quantification of this property is traditionally governed by Ohm's law, expressed as:

$$V = IR \tag{1}$$

Here, V denotes voltage (in joules per coulomb), I is the current (in coulombs per second), and R is the resistance measured in ohms (V/A). The variable R is contingent upon the geometric configuration of the specimen but, for a substantial array of materials, remains invariant with respect to the magnitude of the electrical current.

The attribute of electrical resistivity (ρ) transcends specimen geometry, and correlates with resistance R via the following equation:

$$\rho = \frac{RA}{l} \tag{2}$$

Herein, l denotes the length over which the voltage is gauged, and A stands for the cross-sectional area perpendicular to the flow of current. The units for resistivity are ohm-meters ($\Omega \cdot m$). Employing both Equation 1 and Equation 2, resistivity can be reformulated as follows:

$$\rho = \frac{VA}{Il} \tag{3}$$

Conversely, the property known as electrical conductivity (σ) is used to describe the nature of electrical behavior within materials. Mathematically, it is articulated as the multiplicative inverse of resistivity:

$$\sigma = \frac{1}{\rho} \tag{4}$$

Solid-state materials present a huge range of electrical conductivities, which spans almost 27 orders of magnitude—an unmatched breadth across all known physical properties. This provides the foundation for a taxonomic classification based on electrical conductance, dividing materials into conductors, semiconductors, and insulators [1]. Metals typify good conductors and possess conductivities generally in the realm of 10^7 $(\Omega \cdot m)^{-1}$. Conversely, materials that exhibit electrical conductivities between 10^{-10} and 10^{-20} $(\Omega \cdot m)^{-1}$ are classified as electrical insulators. Intermediately, semiconductors are characterized by electrical conductivities ranging generally from 10^{-6} to 10^4 $(\Omega \cdot m)^{-1}$.

Electronic Conduction

An electrical current emanates from the motion of charge-bearing particles, actuated by an externally imposed electric field. Within this external field, positively charged particles accelerate in alignment with the field vector, while their negatively charged counterparts move in an antiparallel direction. The predominant mechanism underlying current flow in solid materials is the drift of electrons, denoted as electronic conduction. Conversely, in materials exhibiting ionic conduction, a global migratory motion of charged ions contributes to current generation (detailed below).

At ambient temperatures, the vast majority of polymers and ionic ceramics act as electrical insulators. Due to their electronic energy band structures, thermally induced excitations of electrons across the band gap occur infrequently. This accounts for their negligible electronic conductivity values [1].

Various ceramic materials are commonly employed for their insulating capabilities, rendering a high electrical resistivity an advantageous attribute. As thermal conditions escalate, insulating materials manifest an augmentation in electrical conductivity. Intriguingly, this uptick in conductivity may, under certain conditions, surpass that observed in semiconductor materials.

Ionic Conduction

In ionic materials, both cations and anions harbor an electric charge. When subjected to an electric field, these ions are susceptible to diffusive behavior. Consequently, the net displacement of these charged ions culminates in the generation of an electric current, which exists concomitantly with any electron-induced current. The directionalities of anionic and cationic movements are antithetical to each other. The aggregate conductivity of such an ionic material, denoted as σ_{total}, can be mathematically represented as the sum of its electronic and ionic conductivities:

$$\sigma_{total} = \sigma_{electronic} + \sigma_{ionic} \tag{5}$$

These conductive contributions may vary in predominance depending on factors like material composition, purity, and thermal conditions. An associated mobility μ_I can be quantified for each ionic species using the following equation:

$$\mu_I = \frac{n_I e D_I}{kT} \tag{6}$$

Here, n_I and D_I correspond to the ion's valence and diffusion coefficient, respectively, while e is the electron's charge, k the Boltzmann's constant and T is the absolute temperature.

Both the electronic and ionic contributions to conductivity amplify with temperature. However, the majority of ionic materials maintain their insulating characteristics even at elevated temperatures.

Although the conductivity of ions in solid-state materials has been recognized for over two centuries, the phenomenon has gained increased significance in recent years, particularly due to advancements in solid-state fuel cells and batteries. Lithium-ion (Li-ion) batteries operate on the principles of solid-state lithium ion conductivity and are deemed among the most efficient rechargeable power sources for use in mobile electronics and electric cars. Sodium-ion (Na-ion) batteries function in a manner fundamentally analogous to their lithium counterparts, relying on sodium ion conductivity. The primary challenges impeding the progress of both Li-ion and Na-ion batteries include identifying electrolytes with elevated lithium and sodium ion conductivities, as well as engineering electrodes that offer prolonged cycle lifetimes and high reversible specific capacities [2-5].

Detailed explorations into the conductivities of lithium ions, sodium ions, and protons (H^+) in various High-Entropy Oxides can be found in previous chapters of this book.

Dielectricity

A dielectric substance is an electrical insulator (non-metallic) characterized by its potential to display an electric dipole structure. In other words, there is a segregation of positively and negatively charged entities at the molecular or atomic scale.

Due to their interaction with electric fields, dielectrics mainly find utility in capacitors. In the context of a capacitor, applying a voltage results in one plate acquiring a positive charge while the other obtains a negative charge. The consequent electric field aligns itself from the positively charged plate towards the negatively charged one. The capacitance C is defined as the ratio of the charge Q stored on each plate to the voltage V applied across the plates, mathematically represented as follows:

$$C = \frac{Q}{V} \tag{7}$$

The units of capacitance are denoted in coulombs per volt or farads (F).

In the case of a parallel-plate capacitor having a vacuum between the plates, the capacitance can be calculated using the following formula:

$$C = \frac{\epsilon_0 A}{l} \tag{8}$$

where A is the surface area of the plates and l denotes the distance separating them; ϵ_0 is the permittivity of vacuum, a universal constant valued at 8.85×10^{-12} F/m.

If a dielectric medium is placed between the plates, the formula for capacitance becomes:

$$C = \frac{\epsilon A}{l} \tag{9}$$

where ϵ is the permittivity of the dielectric itself, which is always larger than ϵ_0.

The relative permittivity ϵ_r, frequently addressed as the dielectric constant of a material, is defined as the ratio between ϵ and ϵ_0. This value exceeds one and highlights the increase in the charge storage capability upon the inclusion of the dielectric medium between the plates. The dielectric constant serves as a critical material attribute in the design of capacitors.

Ferroelectricity

Materials within the dielectric category known as ferroelectrics manifest spontaneous polarization, meaning they exhibit polarization even in the absence of an external electric field. This is analogous to the permanent magnetism in ferromagnetic materials. For example, in the case of barium titanate, a prototypical ferroelectric material, spontaneous polarization arises due to the geometric arrangement of Ba^{2+}, Ti^{4+} and O^{2-} ions within its unit cell [6]. Hence, a dipole moment forms due to the relative displacements of O^{2-} and Ti^{4+} ions. Above the material's Curie temperature, the unit cell adopts a cubic symmetry, and the ferroelectric properties disappear. Spontaneous polarization is further augmented through interactions between adjacent permanent dipoles, leading them to align collectively.

Piezoelectricity

Certain ceramic materials exhibit piezoelectricity, a phenomenon where an electric field is induced due to mechanical strain upon application of an external force. The directionality of the field is reversible with the inversion of the mechanical force. This property enables these materials to act as transducers, converting electrical energy to mechanical energy and *vice versa*. Initially employed in sonar systems, piezoelectric devices have proliferated into numerous modern applications [7-9]. The piezoelectric properties are predominantly exhibited by materials with complex crystal structures that possess low degrees of symmetry.

Superconductivity

Superconductivity is primarily an electrical property, but its discussion has strong magnetic implications as its predominant application in real life. As high-purity metals approach temperatures close to 0 K, their electrical resistivity decreases steadily to a small but non-zero value specific to the metal. However, certain materials experience a sudden drop in resistivity at extremely low temperatures to almost zero, maintaining this state upon further cooling. These materials are termed superconductors, and the temperature at which this transition occurs is known as the critical temperature [1].

The critical temperature varies among superconductors but typically falls between less than 1 K and around 20 K for metals and metal alloys. However, several studies have shown that complex oxide ceramics (generally cuprates) can have critical temperatures exceeding 100 K [10, 11]. These superconducting ceramics could be promising, but their technological applications are limited by their brittleness.

When the temperature drops below a certain temperature, a sufficiently large applied magnetic field H_C, called the critical field, terminates the superconducting state of a certain material. This critical field is temperature-dependent and decreases as temperature rises. Similarly, a threshold current density J_C exists, below which the material remains superconductive.

The phenomenon of superconductivity has been adequately explained through complex theories [12, 13]. Essentially, paired conducting electrons engage in coordinated movements, making them less susceptible to scattering from thermal vibrations or impurities, thus reducing resistivity to zero. Superconducting materials can be categorized into Type I and Type II based on their magnetic response. Type I superconductors, including elements like aluminum, lead, tin, and mercury, are completely diamagnetic. When the applied magnetic field reaches H_C, the material transitions to normal conduction. Type II superconductors exhibit gradual transitions between superconducting and normal states, defined by lower and upper critical fields H_{C1} and H_{C2}, respectively.

Superconductivity has far-reaching practical applications, such as in MRI machines, high-energy particle accelerators, and magnetically levitated trains [14, 15]. The main challenge remains the necessity for extremely low operating temperatures, which could potentially be mitigated by newly developed materials with higher critical temperatures.

Electrical Properties of High-Entropy Ceramics (HECs)

The presence of multiple cations in High-Entropy Ceramics (HECs) leads to substantial lattice distortions, which have a considerable impact on their electrical properties. Furthermore, the electromagnetic variances among the individual cations and their interplay can contribute to a wide range of intriguing electrical features within HECs. Particularly, many studies have already thoroughly examined the electrical properties of HECs (mainly oxides), with specific emphasis on their electrical conductivity [16-18], piezoelectric features [19, 20] and dielectric properties [21-23].

As previously stated, in the domain of ceramics, electrical conductivity and its inverse, resistivity, garner attention mainly for their potential in creating high-temperature insulators. Owing to the distorted structure and high electron scattering in HEOs, they offer the potential for developing new insulators with very high electrical resistivities. Existing data on the electrical resistivity and conductivity of HEOs are comprehensively summarized in a recent review [24].

Specifically, Lin et al. [25] analyzed the electrical resistivity of (AlCrTaTiZr) O_x thin films fabricated via DC magnetron sputtering, finding that increasing the oxygen concentration led to a rise in electrical resistivity, reaching up to 12 $\mu\Omega$-cm.

Gild et al. [26] measured the electrical conductivity of eight fluorite-structured HEOs, concluding that they exhibited lower electronic conductivities compared to conventional insulating Y_2O_3-stabilized ZrO_2. They also observed that the addition of CaO led to reduced conductivity at low temperatures. Factors such as grain size and grain-boundary resistance were identified as contributing to this reduced conductivity.

Balcerzak et al. [27] studied the rock-salt structured (Co,Cu,Mg,Ni,Zn)O, and reported a peak for its electrical conductivity of 8.039×10^{-2} S cm^{-1} at 1148 K.

Stygar et al. [28] studied the electrical properties of various spinel-structured HEOs and observed high electrical conductivities at high temperatures and a semiconductor behavior at low temperatures, with Seebeck coefficients within the 90-150 μV/K range, which is characteristic of conventional semiconductors.

Conventional materials employed in piezoelectric applications like actuators, sensors, and microelectronics generally rely on the perovskite-structured lead zirconate titanate ($PbTiO_3$) [29, 30]. Nevertheless, the environmental impact of these materials is a concern due to the toxicity of lead. In addition, the high vapor pressure generated during the sintering process of $PbTiO_3$ poses another significant challenge. Therefore, researchers are actively seeking environmentally friendly substitutes for $PbTiO_3$, including for example $Bi_{0.5}Na_{0.5}TiO_3$ [31] or several different HEO systems.

Specifically, Lin et al. [32] introduced a lead-free piezoelectric perovskite-structured system, with the following composition: $(Bi_{1-x-y}Na_{0.925-x-y}Li_{0.075})_{0.5}Ba_xSr_yTiO_3$. Their investigations into its piezoelectric and ferroelectric attributes indicated better piezoelectric coefficients when compared to unmodified $Bi_{0.5}Na_{0.5}TiO_3$.

Zachariasz et al. [33] explored a medium-entropy version of the conventional lead zirconate titanate, i.e. $Pb_{0.94}Sr_{0.06}(Zr_{0.50}Ti_{0.50})_{0.99}Cr_{0.01}O_3$, stating that this material falls under the category of ferroelectric-hard materials, making it suitable for diverse applications such as micromechatronics, resonators, filters, and ultrasonic transducers [34].

Finally, Bochenek et al. [35] synthesized a perovskite-structured HEO, i.e. $(Zr_{0.49}Ti_{0.51})_{0.94}Mn_{0.014}Sb_{0.02}W_{0.014}Ni_{0.02}O_3$, exhibiting impressive piezoelectric attributes, substantial dielectric permittivity values, and low dielectric losses, indicating that such material harbors potential for usage in advanced micromechatronic and microelectronic systems, particularly in actuators.

As for dielectric properties, playing a pivotal role in the conceptualization and fabrication of capacitors used for energy storage, barium titanate ($BaTiO_3$) is the most used material. However, recent research [24] has shown that rock-salt structured High-Entropy Oxides exhibit superior dielectric constants, which can be further optimized by altering the concentration of its elemental components. However, research into the dielectric behavior of HEOs is not merely confined to those with a rock-salt structure; efforts are underway to develop high-entropy dielectric perovskites as well.

Particularly, Zhou et al. [36] synthesized and studied a range of $BaTiO_3$-type perovskites featuring a general formula of $Ba(Zr_{0.2}Ti_{0.2}Sn_{0.2}Hf_{0.2}Me_{0.2})O_3$ (with Me representing either Y, V, Nb, Ta, Mo, or W). Their study revealed excellent stability in permittivity across temperatures ranging from 298 K to 473 K and negligible dielectric losses in a frequency domain of 20 Hz to 2 MHz.

Pu et al. [37] synthesized and studied another perovskite-structured HEO, i.e. $(Na_{0.2}Bi_{0.2}Ba_{0.2}Sr_{0.2}Ca_{0.2})TiO_3$, exhibiting a diffused phase transition and frequency dispersion, boasting an energy density of 1.02 J cm^{-3} at 145 kV cm^{-1}.

Radoń et al. [38] analyzed a novel high-entropy ferrite, i.e. $(Zn_{0.2}Mg_{0.2}Ni_{0.2}Fe_{0.2}Cd_{0.2})Fe_2O_4$, finding that its high-frequency complex dielectric permittivity is temperature and frequency-dependent, and is comparable to commercial $BaTiO_3$.

Finally, regarding superconductivity in High-Entropy ceramics, in a very recent research by Mazza et al. [39], the Ruddlesden–Popper high-entropy oxide series with a generic formula of Me_2CuO_4 (where Me is a combination of 5 elements among La, Pr, Nd, Sm, Eu, Ce, Sr) was studied through charge doping. The obtained results indicate that these materials do not exhibit superconductivity, possibly due to significant distortions in the Cu–O plane, but their study opens new avenues for exploring properties like superconductivity and metal-insulator transitions in HEOs, providing a wider understanding of their unique features.

References

[1] Callister Jr, W.D. & Rethwisch, D.G. (2020). Callister's Materials Science and Engineering. John Wiley & Sons.

[2] Chen, R., Qu, W., Guo, X., Li, L. & Wu, F. (2016). The pursuit of solid-state electrolytes for lithium batteries: From comprehensive insight to emerging horizons. Materials Horizons, 3(6), 487–516.

[3] Goodenough, J.B. & Kim, Y. (2010). Challenges for rechargeable Li batteries. Chemistry of Materials, 22(3), 587–603.

[4] Hou, W., Guo, X., Shen, X., Amine, K., Yu, H. & Lu, J. (2018). Solid electrolytes and interfaces in all-solid-state sodium batteries: Progress and perspective. Nano Energy, 52, 279–291.

[5] Nayak, P.K., Yang, L., Brehm, W. & Adelhelm, P. (2018). From lithium-ion to sodium-ion batteries: Advantages, challenges, and surprises. Angewandte Chemie International Edition, 57(1), 102–120.

[6] Jiang, B., Iocozzia, J., Zhao, L., Zhang, H., Harn, Y.W., Chen, Y. & Lin, Z. (2019). Barium titanate at the nanoscale: Controlled synthesis and dielectric and ferroelectric properties. Chemical Society Reviews, 48(4), 1194–1228.

[7] Tressler, J.F. (2008). Piezoelectric transducer designs for sonar applications. *In:* Piezoelectric and Acoustic Materials for Transducer Applications (pp. 217–239). Boston, MA: Springer US.

[8] Manbachi, A. & Cobbold, R.S. (2011). Development and application of piezoelectric materials for ultrasound generation and detection. Ultrasound, 19(4), 187–196.

[9] Chen, L., Liu, H., Qi, H. & Chen, J. (2022). High-electromechanical performance for high-power piezoelectric applications: Fundamental, progress, and perspective. Progress in Materials Science, 127, 100944.

[10] Capponi, J.J., Chaillout, C., Hewat, A.W., Lejay, P., Marezio, M., Nguyen, N., ... & Tournier, R. (1987). Structure of the 100 K superconductor $Ba_2YCu_3O_7$ between (5÷300) K by neutron powder diffraction. Europhysics Letters, 3(12), 1301.

[11] Ge, J.F., Liu, Z.L., Liu, C., Gao, C.L., Qian, D., Xue, Q.K., ... & Jia, J.F. (2015). Superconductivity above 100 K in single-layer FeSe films on doped $SrTiO_3$. Nature Materials, 14(3), 285–289.

[12] Ginzburg, V.L., Ginzburg, V.L. & Landau, L.D. (2009). On the Theory of Superconductivity (pp. 113–137). Springer Berlin Heidelberg.

[13] Schrieffer, J.R. (2018). Theory of Superconductivity. CRC Press.

[14] Bottura, L., Gourlay, S.A., Yamamoto, A. & Zlobin, A.V. (2015). Superconducting magnets for particle accelerators. IEEE Transactions on Nuclear Science, 63(2), 751–776.

[15] Bernstein, P. & Noudem, J. (2020). Superconducting magnetic levitation: Principle, materials, physics and models. Superconductor Science and Technology, 33(3), 033001.

[16] Li, H., Zhou, Y., Liang, Z., Ning, H., Fu, X., Xu, Z., ... & Peng, J. (2021). High-entropy oxides: Advanced Research on Electrical Properties. Coatings, 11(6), 628.

[17] Wang, Y., Jin, Y.J., Ding, Z.Y., Cao, G., Liu, Z.G., Wei, T., ... & Wang, Y.J. (2022). Microstructure and electrical properties of new high-entropy rare-earth zirconates. Journal of Alloys and Compounds, 906, 164331.

[18] Sun, W., Zhang, F., Zhang, X., Shi, T., Li, J., Bai, Y., ... & Wang, Z. (2022). Enhanced electrical properties of $(Bi_{0.2}Na_{0.2}Ba_{0.2}Ca_{0.2}Sr_{0.2})$ TiO_3 high-entropy ceramics prepared by hydrothermal method. Ceramics International, 48(13), 19492-19500.

[19] Zhang, S. (2023). High entropy design: A new pathway to promote the piezoelectricity and dielectric energy storage in perovskite oxides. Microstructures, 3(1), 2023003.

[20] Sharma, Y., Lee, M.C., Pitike, K.C., Mishra, K.K., Zheng, Q., Gao, X., ... & Ward, T.Z. (2022). High entropy oxide relaxor ferroelectrics. ACS Applied Materials & Interfaces, 14(9), 11962–11970.

[21] Liu, W., Li, F., Chen, G., Li, G., Shi, H., Li, L., ... & Wang, C. (2021). Comparative study of phase structure, dielectric properties and electrocaloric effect in novel high-entropy ceramics. Journal of Materials Science, 56, 18417–18429.

[22] Zhou, S., Pu, Y., Zhang, Q., Shi, R., Guo, X., Wang, W., ... & Ouyang, T. (2020). Microstructure and dielectric properties of high entropy Ba $(Zr_{0.2}Ti_{0.2}Sn_{0.2}Hf_{0.2}Me_{0.2})$ O_3 perovskite oxides. Ceramics International, 46(6), 7430–7437.

[23] Pu, Y., Zhang, Q., Li, R., Chen, M., Du, X. & Zhou, S. (2019). Dielectric properties and electrocaloric effect of high-entropy $(Na_{0.2}Bi_{0.2}Ba_{0.2}Sr_{0.2}Ca_{0.2})$ TiO_3 ceramic. Applied Physics Letters, 115(22).

[24] Akrami, S., Edalati, P., Fuji, M. & Edalati, K. (2021). High-entropy ceramics: Review of principles, production and applications. Materials Science and Engineering: R: Reports, 146, 100644.

[25] Lin, M.I., Tsai, M.H., Shen, W.J. & Yeh, J.W. (2010). Evolution of structure and properties of multi-component (AlCrTaTiZr) Ox films. Thin Solid Films, 518(10), 2732–2737.

[26] Gild, J., Samiee, M., Braun, J.L., Harrington, T., Vega, H., Hopkins, P.E., ... & Luo, J. (2018). High-entropy fluorite oxides. Journal of the European Ceramic Society, 38(10), 3578–3584.

[27] Balcerzak, M., Kawamura, K., Bobrowski, R., Rutkowski, P. & Brylewski, T. (2019). Mechanochemical synthesis of (Co, Cu, Mg, Ni, Zn) O high-entropy oxide and its physicochemical properties. Journal of Electronic Materials, 48, 7105–7113.

[28] Stygar, M., Dąbrowa, J., Moździerz, M., Zajusz, M., Skubida, W., Mroczka, K., ... & Danielewski, M. (2020). Formation and properties of high entropy oxides in co-cr-fe-mg-mn-ni-o system: Novel (cr, fe, mg, mn, ni) 3o4 and (co, cr, fe, mg, mn) 3o4 high entropy spinels. Journal of the European Ceramic Society, 40(4), 1644–1650.

[29] Ye, C., Tamagawa, T., Schiller, P. & Polla, D.L. (1992). Pyroelectric $PbTiO_3$ thin films for microsensor applications. Sensors and Actuators A: Physical, 35(1), 77–83.

[30] Ma, X.H., Li, H.Y., Kweon, S.H., Jeong, S.Y., Lee, J.H. & Nahm, S. (2019). Highly sensitive and selective $PbTiO_3$ gas sensors with negligible humidity interference in ambient atmosphere. ACS Applied Materials & Interfaces, 11(5), 5240–5246.

[31] Lin, D., Kwok, K.W. & Chan, H.L.W. (2008). Structure and electrical properties of $Bi_{0.5}Na_{0.5}TiO_3$–$BaTiO_3$–$Bi_{0.5}Li_{0.5}TiO_3$ lead-free piezoelectric ceramics. Solid State Ionics, 178(37–38), 1930–1937.

[32] Lin, D. & Kwok, K.W. (2009). Structure, ferroelectric and piezoelectric properties of $(Bi1-x-yNa_{0.925}-x-yLi_{0.075})_{0.5}$ $BaxSryTiO_3$ lead-free piezoelectric ceramics. Current Applied Physics, 9(6), 1369–1374.

[33] Zachariasz, R. & Bochenek, D. (2015). Modified PZT ceramics as a material that can be used in micromechatronics. The European Physical Journal B, 88, 1–4.

[34] Setter, N., Damjanovic, D., Eng, L., Fox, G., Gevorgian, S., Hong, S., ... & Streiffer, S. (2006). Ferroelectric thin films: Review of materials, properties, and applications. Journal of Applied Physics, 100(5).

[35] Bochenek, D., Niemiec, P., Skulski, R. & Adamczyk-Habrajska, M. (2019). Electrophysical properties of a multicomponent PZT-type ceramics for actuator applications. Journal of Physics and Chemistry of Solids, 133, 128–134.

[36] Zhou, S., Pu, Y., Zhang, Q., Shi, R., Guo, X., Wang, W., ... & Ouyang, T. (2020). Microstructure and dielectric properties of high entropy Ba $(Zr_{0.2}Ti_{0.2}Sn_{0.2}Hf_{0.2}Me_{0.2})$ O_3 perovskite oxides. Ceramics International, 46(6), 7430–7437.

[37] Pu, Y., Zhang, Q., Li, R., Chen, M., Du, X. & Zhou, S. (2019). Dielectric properties and electrocaloric effect of high-entropy $(Na_{0.2}Bi_{0.2}Ba_{0.2}Sr_{0.2}Ca_{0.2})TiO_3$ ceramic. Applied Physics Letters, 115(22).

[38] Radoń, A., Hawełek, Ł., Łukowiec, D., Kubacki, J. & Włodarczyk, P. (2019). Dielectric and electromagnetic interference shielding properties of high entropy (Zn, Fe, Ni, Mg, Cd) Fe_2O_4 ferrite. Scientific Reports, 9(1), 20078.

[39] Mazza, A.R., Gao, X., Rossi, D.J., Musico, B.L., Valentine, T.W., Kennedy, Z., ... & Ward, T.Z. (2022). Searching for superconductivity in high entropy oxide Ruddlesden–Popper cuprate films. Journal of Vacuum Science & Technology A, 40(1).

Noticeable High-Entropy Ceramics' Magnetic Properties

Magnetic Properties of Materials

Magnetism, i.e. a force that produces either attraction or repulsion between substances, has been recognized for millennia. Nevertheless, the intricate principles that account for these magnetic behaviors have only been explained in more recent history. A multitude of contemporary technological applications depend on magnetism and materials with magnetic properties, such as generators and transformers for electrical power, electric motors, and various communication and entertainment devices like radios, televisions, telephones, computers, as well as audio and video system components.

Materials like iron, certain types of steel, and the naturally-occurring mineral called lodestone, are commonly understood to possess magnetic properties. What may not be as widely recognized is that all materials experience some level of interaction when exposed to a magnetic field. This section offers a concise overview of the genesis of magnetic fields and delves into different categories of magnetism, such as diamagnetism, paramagnetism, ferromagnetism, and ferrimagnetism.

Magnetic Dipoles

Magnetic forces arise from the movement of electrically charged particles and are supplementary to any existing electrostatic forces. It is often useful to conceptualize these forces within the framework of magnetic fields. Hypothetical lines of force can be illustrated to denote the directionality of the force near the origin of the field.

Magnetic dipoles exist within materials that are magnetic and share similarities with electric dipoles [1]. One can envision magnetic dipoles as miniature bar magnets containing north and south poles, rather than as clusters of positive and negative electric charges. In this context, magnetic dipole moments can be symbolized by arrows. The behavior of magnetic dipoles within a magnetic field parallels how electric dipoles react in an electric field.

When subjected to a magnetic field, the field's force exerts a torque, which inclines the dipoles to align with the orientation of the field. A case in point is how a magnetic compass needle synchronizes with Earth's magnetic field.

Magnetic Field Vectors

Prior to exploring the genesis of magnetic moments in solid-state materials, magnetic activity using multiple field vectors has to be firstly outlined. The magnetic field applied externally, often termed the magnetic field strength, is represented by the symbol H. In instances where this magnetic field is created via a cylindrical coil or solenoid—with N turns closely spaced over a length l and a current of magnitude I—the field strength H can be expressed as:

$$H = \frac{N I}{l} \tag{1}$$

The units for H are either ampere-turns per meter or simply amperes per meter.

The concept of magnetic induction or magnetic flux density is labeled as B, indicating the internal field strength of a material exposed to a H field. The units for B are teslas or webers per square meter (Wb/m^2). Both B and H are vector quantities, defined by their magnitude and spatial direction.

The relationship between magnetic field strength and flux density can be denoted as:

$$B = \mu H \tag{2}$$

Here, μ denotes the permeability, a characteristic of the specific medium through which the H field travels and where B is measured. The units for permeability are webers per ampere-meter (Wb/A m) or Henries per meter (H/m).

In a vacuum setting, Equation (2) becomes:

$$B_0 = \mu_0 H \tag{3}$$

Here, μ_0 represents the permeability of a vacuum, a universal constant valued at $1.257 \cdot 10^{-6}$ H/m, while the parameter B_0 denotes the flux density within a vacuum.

Various parameters can be employed to describe the magnetic characteristics of solids. One such parameter is the relative permeability, μ_R, defined as:

$$\mu_R = \frac{\mu}{\mu_0} \tag{4}$$

The relative permeability, which is dimensionless, quantifies the extent to which a material can be magnetized or how effortlessly a B field can be generated under the influence of an external H field.

Another field parameter, denoted by M and termed the magnetization of a certain material, can be expressed as follows:

$$B = \mu_0 H + \mu_0 M \tag{5}$$

When subjected to a *H* field, the magnetic moments within a material align with the field, reinforcing its effect. The term $\mu_0 M$ in Equation (5) quantifies this influence.

The magnitude of *M* is proportional to the applied field, expressed as:

$$M = \chi_m H \tag{6}$$

In this equation, χ_m is known as the magnetic susceptibility, a dimensionless quantity. Magnetic susceptibility and relative permeability are interrelated according to:

$$\chi_m = \mu_r - 1 \tag{7}$$

Each magnetic field parameter has a dielectric counterpart. For instance, the *B* and *H* fields correspond to the dielectric displacement *D* and the electric field $\mathcal{E}$, respectively, while permeability μ is analogous to permittivity ϵ. Additionally, magnetization *M* and polarization *P* are comparable.

Diamagnetism and Paramagnetism

Diamagnetism is a weak and temporary magnetic property that endures only as long as an external magnetic field is applied. It originates from alterations in the orbital movements of electrons caused by the presence of an external magnetic field. The resultant magnetic moment is minimum and opposes the direction of the applied magnetic field. Consequently, the relative permeability (μ_r) is slightly less than one, and the magnetic susceptibility is negative. This means that the internal magnetic flux density (*B*) in a diamagnetic material is weaker than it would be in a vacuum. The volume susceptibility (χ_m) for diamagnetic substances is approximately -10^{-5}. When exposed to a powerful electromagnet, these materials tend to move towards areas where the magnetic field is weaker.

Figure 1 schematically depicts the magnetic dipoles in a diamagnetic material, both in the presence and absence of an external field [2]. Atomic dipole moments are represented by arrows, different from electron moments.

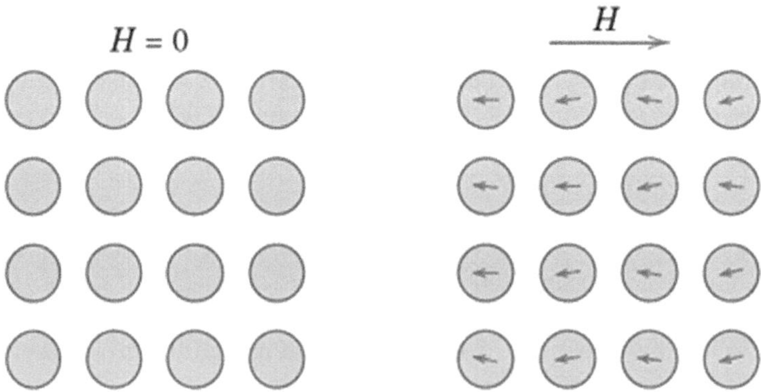

Fig. 1: The atomic dipole configuration for a diamagnetic material with and without a magnetic field (taken from [2]).

Diamagnetism exists in all materials but is so feeble that it is noticeable only when other forms of magnetism are entirely absent. This type of magnetism has limited practical applications.

In contrast to diamagnetism, some solid materials have atoms with permanent dipole moments arising from incomplete cancelation of electron spin and orbital magnetic moments. In the absence of an external magnetic field, these atomic magnetic moments are randomly oriented, resulting in no net macroscopic magnetization. However, these atomic dipoles can align preferentially with an external magnetic field, a phenomenon known as paramagnetism (illustrated in Figure 2).

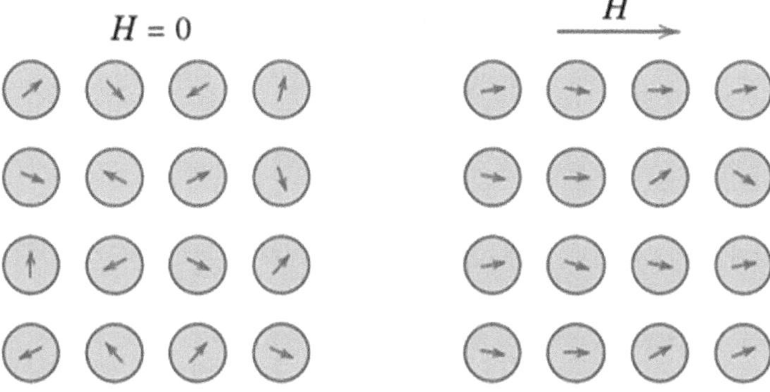

Fig. 2: The atomic dipole configuration for a paramagnetic material with and without a magnetic field (taken from [2]).

These dipoles interact individually, without any mutual influence. Their alignment augments the external magnetic field, leading to a $\mu_r > 1$ and a positive, albeit small, magnetic susceptibility. Susceptibilities for paramagnetic materials vary between approximately 10^{-5} and 10^{-2}.

Both diamagnetic and paramagnetic substances are classified as non-magnetic because they display magnetization only when subjected to an external magnetic field. In both cases, the internal flux density B is nearly identical to what it would be in a vacuum.

Ferromagnetism

In a distinct category, certain metals possess intrinsic magnetic moments even without an external magnetic field and exhibit strong and stable magnetization. This is the hallmark of ferromagnetism, seen in transition metals like body-centered cubic (α-ferrite) iron, cobalt, nickel, and some rare-earth metals like gadolinium (Gd). These materials can achieve magnetic susceptibilities as high as 10^6. Consequently, H is much smaller than M, and from Equation (5), it is possible to deduce:

$$B \approx \mu_0 M \tag{8}$$

Permanent magnetic moments in ferromagnetic materials stem from atomic magnetic moments due to non-canceling electron spins, as determined by the electronic structure. Orbital magnetic moments also contribute, but are relatively insignificant. In these materials, interactions between neighboring atomic magnetic moments cause them to align, even without an external field, as illustrated schematically in Figure 3.

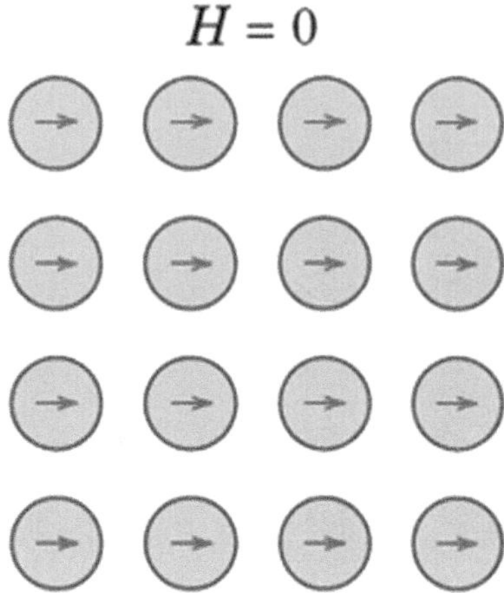

Fig. 3: The atomic dipole configuration for a ferromagnetic material (taken from [2]).

While the exact origin of these coupling forces remains unclear, they are presumed to be a result of the material's electronic structure. Such alignment occurs over large crystal domains.

The highest achievable magnetization, known as saturation magnetization (M_s), is the magnetization when all magnetic dipoles within the material align with an external field. A corresponding saturation flux density (B_s) also exists. The saturation magnetization is the product of the net magnetic moment per atom and the total number of atoms.

Antiferromagnetism

In materials other than ferromagnetic ones, coupling of magnetic moments between neighboring atoms or ions can result in an antiparallel alignment, known as antiferromagnetism. One example is manganese oxide (MnO), a ceramic and ionic compound. The oxygen ions (O^{2-}) do not contribute to the magnetic moment as they display a complete cancellation of both spin and orbital moments. On the other hand, Mn^{2+} ions do have a net magnetic moment mainly originating from their spin. The adjacent Mn^{2+} ions align in such a way that their moments are oppositely directed, cancelling each other out and resulting in no net magnetism for the material as a whole.

Ferrimagnetism

Certain ceramic materials also display a type of permanent magnetization called ferrimagnetism. Although the macroscopic characteristics of ferrimagnetic and ferromagnetic materials are akin, the difference lies in the source of their net magnetic moments. As an example, cubic ferrites (i.e. ionic compounds that can be described by the chemical formula MFe_2O_4) can be considered. The magnetic moments of Fe ions in these materials also interact antiparallelly like in antiferromagnetic materials. However, a net ferrimagnetic moment arises due to the incomplete negation of these moments. In cubic ferrites, Fe^{2+} ions contribute to the net magnetization of the material.

Different compositions of cubic ferrites can be synthesized by substituting other divalent metal ions for some of the iron ions. This allows for a variation in the magnetic properties of these compounds. In addition, other ceramic materials like hexagonal ferrites and garnets also display ferrimagnetic properties [3-6]. Hexagonal ferrites have a different crystal structure and chemical formula, while garnets possess an even more intricate crystal configuration.

Ferrimagnetic materials don't reach as high levels of saturation magnetization as ferromagnetic materials. Nevertheless, since they are ceramic, they are excellent electrical insulators, which is a desirable trait for applications such as high-frequency transformers [2].

Curie Temperature

The term "Curie temperature" (T_C) refers to the critical temperature above which a ferromagnetic or ferrimagnetic material loses its magnetic ordering and transitions into a paramagnetic state. In other words, below the Curie temperature, the magnetic moments of a certain material are aligned in a stable manner, resulting in a net magnetization. Above this temperature, thermal agitation becomes strong enough to disrupt this alignment, causing the material to lose its spontaneous magnetization.

The Curie temperature is specific to each material and depends on its crystal structure, the types of atoms involved, and the strength of the exchange interactions among them. In some materials, the Curie temperature may differ in different crystallographic directions, which is known as magnetic anisotropy [7, 8].

The relationship between magnetization (M), temperature (T), and Curie temperature (T_C) is often described mathematically (near the Curie temperature) by the Curie-Weiss Law:

$$M = \frac{C(T - T_c)}{T} \tag{9}$$

where C is the Curie constant.

The Curie constant is a material-specific parameter that measures a material's magnetic susceptibility and is determined by the magnetic moment of the individual atoms, ions, or molecules that make up the material. It depends on the number of magnetic moments per unit volume and their intrinsic properties. The Curie constant is often experimentally determined and can be used to understand the magnetic properties of the material at a microscopic level.

The units of the Curie constant depend on the units used for magnetization, magnetic field, and temperature. In the SI system, the Curie constant typically has units of m^3/K or J/K per mole of magnetic ions.

Magnetic Properties of High-Entropy Ceramics

The magnetic characteristics of oxide materials are intricately shaped by several factors including their chemical composition, their structural framework, and the states of their spin-electronics [9, 10]. When it comes to assessing the magnetic behaviors of high-entropy oxides (HEOs), the task is particularly complex. This complexity arises from the lattice anisotropy, inherently disordered chemical compositions, structural irregularities, and the stochastic distribution of magnetically active atoms, each with unique electronic and ionic configurations [11].

The varied neighboring ionic arrangements in HEOs contribute to a highly complex magneto-electronic free-energy landscape, and in turn, stabilize unconventional spin-electronic configurations [12]. Numerous scholarly publications have subsequently reviewed the magnetic characteristics inherent to HEOs.

In one such work, Meisenheimer et al. [13] experimented with $(Mg_{0.2}Co_{0.2}Ni_{0.2}Cu_{0.2}Zn_{0.2})O$ and succeeded in inducing an antiferromagnetic order within a heterostructured MgO-HEO composite. They revealed that at the boundary between the ferromagnetic and antiferromagnetic regions, it is possible to modulate exchange coupling through careful adjustment of the disordered chemical composition and the quantity of magnetic ions. Their findings indicate that high levels of chemical disorder are present at equiatomic concentrations, and these levels are amplified when cobalt concentrations are increased. They further demonstrated that enhancing cobalt concentration in HEOs could surpass the exchange bias values observed in traditional permalloy/CoO heterostructures [14].

Zhang et al. [15] studied the magnetic structure of $(Mg_{0.2}Co_{0.2}Ni_{0.2}Cu_{0.2}Zn_{0.2})O$ employing a combination of direct and alternating current magnetic susceptibility assays along with elastic and inelastic neutron scattering techniques. Their analysis indicated that the material undergoes a slow magnetic transformation to achieve a long-range antiferromagnetic ground state, which becomes evident below the Curie temperature of 113 K. Furthermore, their research revealed that within this magnetic structure, ferromagnetic (111) planes exist where electron spins are oriented antiparallel between neighboring atomic planes [15].

Mao et al. [16] revealed that the rock-salt $(Mg_{0.2}Co_{0.2}Ni_{0.2}Cu_{0.2}Zn_{0.2})O$ compound, produced through solution combustion synthesis, possesses antiferromagnetic long-range ordering characteristics below its Curie temperature of 106 K and shows paramagnetic behaviors at room temperature. These features could be attributed to the presence of super-exchange interactions within the high-entropy oxide. Moreover, from such study it is possible to say that the Curie temperature for HEOs is lower than those for their binary oxide counterparts, such as CoO, NiO, and CuO in the case of [16], owing to the disordered chemical compositions and the significant proportions of non-magnetic atoms in HEOs.

Further studies on the rock-salt $(Mg_{0.2}Co_{0.2}Ni_{0.2}Cu_{0.2}Zn_{0.2})O$ system [17] indicated that it attains long-range magnetic ordering at temperatures below 120 K, despite the inherent disorder in the distribution of its magnetic ions. Additionally, it has been demonstrated that the magnetic ground states in rocksalt HEOs could be manipulated through compositional substitutions [17].

A subsequent study authored by Witte et al. [18] demonstrated that for various high-entropy perovskites, such as $Y(Co_{0.2}Cr_{0.2}Fe_{0.2}Mn_{0.2}Ni_{0.2})O_3$ and $La(Co_{0.2}Cr_{0.2}Fe_{0.2}Mn_{0.2}Ni_{0.2})O_3$, predominantly antiferromagnetic behaviors with minor ferromagnetic contributions are observed, which they attributed to the existence of ferromagnetic clusters embedded within an antiferromagnetic lattice. In a follow-up study [19], the same authors employed X-ray absorption near edge spectroscopy, magnetometry, and Mössbauer spectroscopy to investigate the magnetic properties of different additional high-entropy perovskite systems. Despite the chemical disorder in rare-earth sites, the magnetic characteristics of HEOs exhibit similarities to their binary oxide equivalents but with unique magnetic features such as noncollinearity, spin reorientation transitions, and high coercive fields reaching up to 2 tesla at room temperature.

Vinnik et al. [20] studied the extremely complex high-entropy oxide $Ba(Fe_{5.83}Al_{1.19}Ti_{1.08}Cr_{1.12}Cu_{0.78}Ga_{1.03}In_{0.97})O_{19}$ owning a magnetoplumbite, demonstrating that it exhibits ferromagnetic properties at low temperatures around 50 K and transitions to a paramagnetic state at ambient conditions. The relatively low observed Curie temperature of 180.4 K for such system can be attributed to the dilution of high magnetic moment ions by ions with zero magnetic moments. Furthermore, Vinnik et al.'s study suggests that increasing the concentration of high-spin cations could enhance the paramagnetic attributes of HEOs.

Regarding spinel structured HEOs, Mao et al. [21, 22] observed long-range ferromagnetic behavior below their respective Curie temperatures, mainly due to elevated super-exchange interactions stemming from chemical disordering. Replacement of Co^{2+} or Ni^{2+} by Zn^{2+}, a nonmagnetic ion, weakened the ferromagnetic ordering and overall magnetic moments, thereby offering an effective approach for tailoring magnetic properties.

In conclusion, the magnetic properties of high-entropy oxides present a complex but promising landscape for future research and technological applications. The multifaceted interplay between chemical disorder, cation occupation, and oxidation states in these systems complicates the straightforward understanding of their magnetic behaviors. While initial studies have offered crucial insights into some HEO systems, a comprehensive understanding is far from being complete. However, research efforts on HEOs magnetic properties have the potential to unlock conventionally inaccessible magneto-electric free-energy minima, thus offering novel avenues for material engineering. Furthermore, a detailed mapping of structural-magnetic phase diagrams as a function of composition or strain will provide avenues to reversibly control the magnetic features of HEOs.

In the expansive composition space of high-entropy oxides, their magnetic properties are inextricably linked with electric and optical features, possibly offering a fertile ground for the discovery of new physical phenomena too.

References

[1] Reis, M. (2013). Fundamentals of Magnetism. Elsevier.

[2] Callister Jr, W.D. & Rethwisch, D.G. (2020). Fundamentals of Materials Science and Engineering: An Integrated Approach. John Wiley & Sons.

[3] Kim, S.K., Beach, G.S., Lee, K.J., Ono, T., Rasing, T. & Yang, H. (2022). Ferrimagnetic spintronics. Nature Materials, 21(1), 24-34.

[4] Liu, W., Cheng, B., Ren, S., Huang, W., Xie, J., Zhou, G., ... & Hu, J. (2020). Thermally assisted magnetization control and switching of $Dy_3Fe_5O_{12}$ and $Tb_3Fe_5O_{12}$ ferrimagnetic garnet by low density current. Journal of Magnetism and Magnetic Materials, 507, 166804.

[5] Wu, C., Wang, W., Li, Q., Wei, M., Luo, Q., Fan, Y., ... & Yu, Z. (2022). Barium hexaferrites with narrow ferrimagnetic resonance linewidth tailored by site-controlled Cu doping. Journal of the American Ceramic Society, 105(12), 7492–7501.

[6] Emori, S. & Li, P. (2021). Ferrimagnetic insulators for spintronics: Beyond garnets. Journal of Applied Physics, 129(2).

[7] Han, R., Jiang, Z. & Yan, Y. (2020). Prediction of novel 2D intrinsic ferromagnetic materials with high Curie temperature and large perpendicular magnetic anisotropy. The Journal of Physical Chemistry C, 124(14), 7956–7964.

[8] Liu, W., Tong, J., Deng, L., Yang, B., Xie, G., Qin, G., ... & Zhang, X. (2021). Two-dimensional ferromagnetic semiconductors of rare-earth monolayer GdX2 (X = Cl, Br, I) with large perpendicular magnetic anisotropy and high Curie temperature. Materials Today Physics, 21, 100514.

[9] Bhattacharya, A. & May, S.J. (2014). Magnetic oxide heterostructures. Annual Review of Materials Research, 44, 65–90.

[10] Choi, K.J., Biegalski, M., Li, Y.L., Sharan, A., Schubert, J., Uecker, R., ... & Eom, C.B. (2004). Enhancement of ferroelectricity in strained $BaTiO_3$ thin films. Science, 306(5698), 1005–1009.

[11] Sharma, Y., Zheng, Q., Mazza, A.R., Skoropata, E., Heitmann, T., Gai, Z. et al. (2020). Magnetic anisotropy in single-crystal high-entropy perovskite oxide $La(Cr_{0.2}Mn_{0.2}Fe_{0.2}Co_{0.2}Ni_{0.2})O_3$ films, Physical Review Materials, 4, 014404.

[12] Sarkar, A., Kruk, R. & Hahn, H. (2021). Magnetic properties of high entropy oxides. Dalton Transactions, 50(6), 1973–1982.

[13] Meisenheimer, P.B., Kratofil, T.J. & Heron, J.T. (2017). Giant enhancement of exchange coupling in entropy-stabilized oxide heterostructures. Scientific Reports, 7, 13344.

[14] Oses, C., Toher, C. & Curtarolo, S. (2020). High-entropy ceramics. Nature Reviews Materials, 5(4), 295–309.

[15] Zhang, J., Yan, J., Calder, S., Zheng, Q., McGuire, M.A., Abernathy, D.L., ... & Hermann, R.P. (2019). Long-range antiferromagnetic order in a rocksalt high entropy oxide. Chemistry of Materials, 31(10), 3705–3711.

[16] Mao, A., Xiang, H.Z., Zhang, Z.G., Kuramoto, K., Yu, H. & Ran, S. (2019). Solution combustion synthesis and magnetic property of rock-salt $(Co_{0.2}Cu_{0.2}Mg_{0.2}Ni_{0.2}Zn_{0.2})$ O high-entropy oxide nanocrystalline powder. Journal of Magnetism and Magnetic Materials, 484, 245–252.

[17] Jimenez-Segura, M.P., Takayama, T., Bérardan, D., Hoser, A., Reehuis, M., Takagi, H. & Dragoe, N. (2019). Long-range magnetic ordering in rocksalt-type high-entropy oxides. Applied Physics Letters, 114(12).

[18] Witte, R., Sarkar, A., Kruk, R., Eggert, B., Brand, R.A., Wende, H. & Hahn, H. (2019). High-entropy oxides: An emerging prospect for magnetic rare-earth transition metal perovskites. Physical Review Materials, 3(3), 034406.

[19] Witte, R., Sarkar, A., Velasco, L., Kruk, R., Brand, R.A., Eggert, B., ... & Hahn, H. (2020). Magnetic properties of rare-earth and transition metal based perovskite type high entropy oxides. Journal of Applied Physics, 127(18).

[20] Vinnik, D.A., Trofimov, E.A., Zhivulin, V.E., Zaitseva, O.V., Zherebtsov, D.A., Starikov, A.Y. & Taskaev, S.V. (2020). The new extremely substituted high entropy (Ba, Sr, Ca, La)Fe6-x (Al, Ti, Cr, Ga, In, Cu, W)xO_{19} microcrystals with magnetoplumbite structure. Ceramics International, 46(7), 9656–9660.

[21] Mao, A., Xie, H.X., Xiang, H.Z., Zhang, Z.G., Zhang, H. & Ran, S. (2020). A novel six-component spinel-structure high-entropy oxide with ferrimagnetic property. Journal of Magnetism and Magnetic Materials, 503, 166594.

[22] Mao, A., Xiang, H.Z., Zhang, Z.G., Kuramoto, K., Zhang, H. & Jia, Y. (2020). A new class of spinel high-entropy oxides with controllable magnetic properties. Journal of Magnetism and Magnetic Materials, 497, 165884.

Noticeable High-Entropy Ceramics' Optical Properties

Optical Properties of Materials

Optical properties pertain to how a substance reacts when subjected to electromagnetic radiation, especially visible light.

Traditionally, electromagnetic radiation is seen as wave-based, comprising perpendicular electric and magnetic field components, also orthogonal to the wave's propagation direction. Different types of electromagnetic radiation include light, radiant heat, radar, radio waves, and x-rays. Each type is mainly defined by a specific wavelength range and the manner in which it is produced. The spectrum of electromagnetic radiation extends from gamma rays (emitted by radioactive substances) with wavelengths in the order of 10^{-12} m (10^{-3} nm) to x-rays, ultraviolet, visible, and infrared light, and culminating in radio waves with wavelengths up to 10^{5} m. This spectrum is logarithmically plotted in Figure 1 [1].

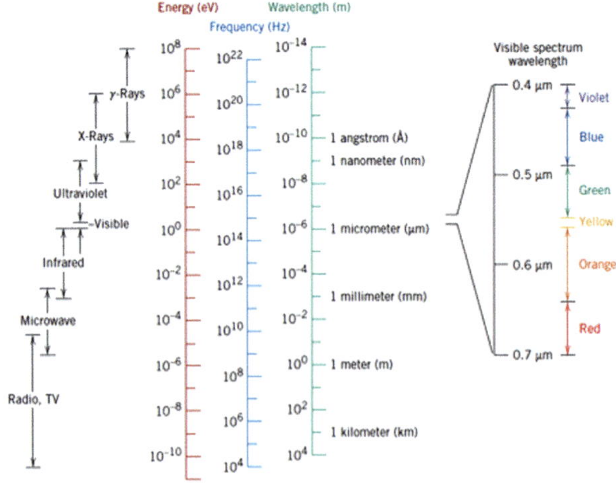

Fig. 1: The spectrum of electromagnetic radiation. Taken from [1].

Visible light occupies a rather confined section of this spectrum, with wavelengths spanning approximately from 0.4 to 0.7 micrometers (μm). Wavelength determines the color we perceive; for instance, violet light has a wavelength around 0.4 μm, while green and red wavelengths are approximately 0.5 and 0.65 μm, respectively. White light is essentially a composite of all colors.

In a vacuum, all forms of electromagnetic radiation travel at a constant speed, denoted by c, which is 3×10^8 m/s. This speed c is correlated to the electric permittivity (ϵ_0) and the magnetic permeability (μ_0) of a vacuum, through the following equation:

$$c = \frac{1}{\sqrt{\epsilon_0 \mu_0}} \tag{1}$$

Moreover, the frequency (ν) and the wavelength (λ) of electromagnetic radiation are governed by their velocity, expressed by the following equation:

$$c = \lambda \nu \tag{2}$$

Frequency is measured in hertz (Hz), with 1 Hz equaling one oscillation per second.

From a quantum-mechanical viewpoint, electromagnetic radiation is composed of discrete energy packets termed photons. The energy (E) of a photon is quantized and can only assume specific values, governed by the following equation:

$$E = h\nu = \frac{hc}{\lambda} \tag{3}$$

where h is Planck's constant, valued at 6.63×10^{-34} J·s. Consequently, photon energy is directly proportional to radiation frequency and inversely proportional to wavelength.

In contexts concerning optical interactions between radiation and matter, it is often easier to consider light as a photon ensemble. In all other cases, wave-based descriptions are generally more suitable.

Refraction

When light passes through the bulk of transparent substances, its speed diminishes and, consequently, the light path deviates at the boundary of the material. This phenomenon is known as refraction.

The refractive index n is calculated as the quotient of the speed of light in a vacuum c over the velocity within the material v, expressed as:

$$n = \frac{c}{v} \tag{4}$$

The value of n, or the extent to which the light is bent, is contingent on the light's wavelength.

This phenomenon is visually exemplified by the dispersion of white light into its constituent colors when passed through a glass prism, as depicted in an adjacent photograph. Each color experiences differing degrees of deflection as it enters and exits the glass, resulting in the partitioning of colors. The refractive index not only governs the trajectory of light but also has implications on the portion of light that gets reflected at the surface.

As Equation (1) outlines the speed of light c, a parallel formula gives the velocity v of light within a certain substance as:

$$v = \frac{1}{\sqrt{\epsilon\mu}} \qquad (5)$$

where ϵ and μ stand for the permittivity and permeability of the given substance, respectively. Combining Equation (5) and Equation (4):

$$n = \sqrt{\epsilon_r\mu_r} \qquad (6)$$

Here, ϵ_r and μ_r represent relative permittivity and relative permeability. Because most substances exhibit negligible magnetism, $\mu_r \approx 1$, leading to:

$$n \approx \sqrt{\epsilon_r} \qquad (7)$$

For transparent substances, the refractive index is linked to the dielectric constant. The occurrence of refraction is connected to electronic polarization, especially at higher frequencies pertinent to visible light. Consequently, the dielectric constant's electronic component can be derived from measurements of the refractive index using Equation (7).

Due to electronic polarization causing the slowing down of electromagnetic radiation within a substance, the dimensions of the constituent atoms or ions forming that substance significantly influence this effect. Generally, the larger an atom or ion, the more pronounced is the electronic polarization, leading to a lower speed and a higher refractive index. For standard soda-lime glass, the refractive index is roughly 1.5. Introducing large ions like barium and lead (as BaO and PbO) can substantially elevate the glass refractive index. For instance, glasses containing 90 wt% PbO exhibit a refractive index of around 2.1 [1].

In crystalline ceramics with cubic structures and glasses, the refractive index is directionally independent, making it isotropic. In contrast, non-cubic crystals demonstrate an anisotropic refractive index, where the index is highest along directions with the greatest ionic density.

Reflection

As light transitions from one material to another with a dissimilar refractive index, a portion of the light disperses at the boundary of the two substances, regardless of their transparency. The parameter 'Reflectivity' (R) quantifies the percentage of incoming light that gets reflected back at the junction. Mathematically, reflectivity can be expressed as follows:

$$R = \frac{I_R}{I_0} \qquad (8)$$

where I_0 and I_R denote the intensity levels of the incoming and reflected light beams, respectively.

If the light impacts the interface at a right angle, the reflectivity is given by:

$$R = \left(\frac{n_2 - n_1}{n_2 + n_1} \right)^2 \tag{9}$$

where n_1 and n_2 are the refractive indices of the initial and secondary media. If the incoming light is not perpendicular to the interface, then the reflectivity depends on the angle of entry. When light crosses from a vacuum or air into a solid medium, the reflectivity can be represented as:

$$R = \left(\frac{n_s - 1}{n_s + 1} \right)^2 \tag{10}$$

Equation (10) considers air's refractive index that is nearly 1. Consequently, the greater the refractive index of the solid material, the higher will be its reflectivity. In the case of standard silicate glasses, reflectivity stands at around 0.05.

It is worth noting that just like the refractive index of a solid relies on the wavelength of the entering light, reflectivity also varies in relation to the wavelength. In optical equipment like lenses, reflective losses are substantially mitigated by applying ultra-thin coatings of dielectric substances, such as magnesium fluoride (MgF_2) [1].

Absorption

Materials that are not metallic can either block visible light or allow it through; when they are transparent, they often display distinct colors. Light absorption in these substances occurs through two fundamental processes that also dictate their light transmission attributes. The first of these is electronic polarization, which is predominantly significant only around the relaxation frequencies of the constituent atoms of the material. The second process is the transition of electrons from the valence band to the conduction band, a property that is reliant on the electronic band structure of a given material.

A light photon can be absorbed through the elevation or activation of an electron from the almost-filled valence band to a vacant state in the conduction band; this action results in a free electron in the conduction band and a hole in the valence band. The energy required for this activation, represented as ΔE, is connected to the frequency of the absorbed photon by a specific equation. This absorption can occur only if the photon's energy surpasses the energy of the band gap E_g, i.e. when $hv > E$.

The lower limit of the wavelength for visible light is approximately 0.4 μm. Considering that the speed of light c is 3×10^8 m/s and Planck's constant h is 4.13×10^{-15} eV·s, the upper bound for E_g permitting the absorption of visible light can be calculated to be around 3.1 eV.

Materials that have a band gap energy exceeding this value will not absorb visible light and will appear transparent and colorless if they are highly pure. On the other hand, the upper limit of the wavelength for visible light is approximately 0.7 µm; calculating the lower bound for E_g where visible light absorption occurs yields about 1.8 eV. Thus, materials with band gap energies less than this value will absorb all visible light, rendering them opaque. Materials with band gap energies ranging between 1.8 and 3.1 eV will absorb only a part of the visible spectrum and will therefore manifest colors.

Every non-metallic material reaches a point of opacity at some wavelength, and this is dictated by its E_g. For instance, diamond, with a band gap energy of 5.6 eV, blocks any radiation that has a wavelength shorter than approximately 0.22 µm [1].

Transmission

The processes of light absorption, reflection, and transmission can be understood in the context of light traveling through a transparent solid. When a beam with an initial intensity I_0 strikes the front surface of a material with a specific thickness l and absorption coefficient β, the intensity IT of the light that emerges from the back surface can be represented by the following equation:

$$IT = I_0 \left(1 - R\right)^2 e^{-\beta l} \tag{11}$$

Here, R stands for the reflectance. Equation (11) presumes that the medium outside both the front and back surfaces is identical.

Therefore, the portion of the initial light that successfully passes through a transparent object is contingent on the losses due to absorption and reflection. Additionally, the sum of the reflectance R, absorptivity A, and transmissivity T will always equal one. It should be noted that each of these variables (i.e. R, A, and T) is wavelength-dependent. For example, across the visible spectrum for a green-colored glass, at a wavelength of 0.4 µm, the proportions of light that are transmitted, absorbed, and reflected are approximately 0.90, 0.05, and 0.05, respectively. However, when the wavelength changes to 0.55 µm, these proportions alter to about 0.50, 0.48, and 0.02, respectively.

Photocatalysts

The term "photocatalyst" merges two concepts: "photo," pertaining to photons and light, and "catalyst," denoting a substance that modulates the reaction kinetics when present. Consequently, photocatalysts are substances that alter the rate of chemical reactions upon light irradiation, a process termed as photocatalysis. This involves the concurrent action of light and a semiconductor material. The material that both absorbs light and functions as a catalyst in chemical processes is identified as a photocatalyst, which is inherently semiconducting in nature. Within photocatalysis, electron-hole pairs are generated when semiconducting materials are exposed to light.

Photocatalytic reactions are dichotomized based on the phase congruence of the reactants:

- Homogeneous photocatalysis occurring when the semiconductor and reactant exist in the same state, i.e. gas, solid, or liquid.
- Heterogeneous photocatalysis occurring when the semiconductor and reactant exist in different states (for example, the former is solid and the latter is gaseous).

Different materials can be segregated into three fundamental categories based on their band gap E_g (i.e. the energy disparity between the highest occupied molecular orbital, also known as the valence band, and the lowest unoccupied molecular orbital, or the conduction band gap):

- Conductor (metal): $E_g < 1.0$ eV
- Semiconductor: 1.5 eV $\leq E_g \leq 3.0$ eV
- Insulator: $E_g > 5.0$ eV

Semiconductors possess the ability to conduct electricity even at ambient temperatures when illuminated, thereby functioning as photocatalysts. Upon light exposure of a given wavelength (i.e. at adequate energy), an electron from the valence band absorbs photon energy and transitions to the conduction band, consequently leaving a hole in the valence band. This leads to the formation of a photoexcited state, generating an electron-hole pair. The excited electron and the hole are involved in the reduction of an acceptor and oxidation of donor molecules, respectively. Significantly, photocatalysts simultaneously offer oxidative and reductive environments depending on the relative energy levels of the semiconductor's conduction and valence bands and the substrate's redox potential [2].

Photocatalysts find applications in diverse fields such as antifouling [3], antifogging [4], energy conservation and storage [5-7], odor neutralization [8], sterilization [9], air purification [10, 11], and wastewater treatment [12–14]. Owing to their electronic structure, photocatalysts serve as sensitizers in photoredox processes and are capable of complete mineralization of various organic pollutants including aromatic compounds, halogenated hydrocarbons, insecticides, pesticides, dyes, and surfactants [2].

Optical Properties of High-Entropy Ceramics

Oxides exhibit a diverse array of optical properties that are not only theoretically intriguing for their insights into electronic structure but also hold significant promise for future technological applications. Compared to other attributes of High-Entropy Oxides (HEOs) that have been more extensively researched, there is a relative scarcity of studies specifically focusing on their optical properties.

Light absorption and the associated bandgap are critical factors that profoundly influence the functionalities of insulating and semiconducting oxides. The absorption of light is determined not only by the positioning of the valence and conduction bands but also by lattice deformations and defects, notably vacancies. HEOs are unique due to their distorted lattice structures, prompting multiple research initiatives to study their light absorption and bandgap properties in detail.

Notably, Chen et al. [15] explored the optical bandgap of a series of $Al_xCoCrCuFeNi$ oxides (x varying among 0.5, 1, 2), finding it variable and ranging from 1 to 2 eV, and indicating that such HEOs possess commendable transparency in the infrared spectrum (1000-4000 cm^{-1}).

Cheng et al. [16] assessed the influence of pressure up to ~35 GPa on the phase transformation and bandgap fluctuations of a fluorite-structured $(Ce_{0.2}La_{0.2}Pr_{0.2}Sm_{0.2}Y_{0.2})O_{2-\delta}$ HEO, demonstrating that the bandgap and light absorption properties were extremely sensitive to pressure changes for such system. Thus, it seems that a unique pressure-dependent behavior in terms of optical properties differentiates fluorite-structured HEOs from most low-entropy oxides. Additional research on $(Ce_{0.2}La_{0.2}Pr_{0.2}Sm_{0.2}Y_{0.2})O_{2-\delta}$ HEO by Sarkar et al. [17] focused on the fluctuation of the bandgap under varying environmental conditions. Their findings suggested reversible changes in the bandgap due to alterations in the oxidation states and orbital occupancies of specific elements within such high-entropy compound.

Genrally speaking, oxides containing rare-earth elements have been identified as materials able to convert wavelengths through two distinct phenomena [18]: up-conversion, which translates longer light wavelengths to shorter ones, and down-shifting, which accomplishes the opposite by converting shorter wavelengths to longer ones. Given that behavior, Hanioka et al. [19] examined the wavelength conversion properties of a rare-earth-based HEO with the following chemical formula: $Y_{1-x-y}Er_xYb_yBaZn_3AlO_7$, observing a down-shifting in the 400-800 nm range when excited at a wavelength of 365 nm. At a different excitation wavelength of 980 nm, conversely, up-conversion peaks have been observed at specific wavelengths, highlighting the HEO's potential for solar cell applications.

Finally, Zhang et al. [20] fabricated transparent HEOs with potentially beneficial optical properties for applications in optical windows, along with superior refractive indices and reduced wavelength dispersions compared to conventional materials.

Anyway, the intricate relationship between the electronic structure and the optical properties of differently-structured HEOs still offers a fertile ground for further exploration and potential applications in areas ranging from optoelectronics to renewable energy technologies.

Photocatalytic Properties of High-Entropy Ceramics

Driven by their unique optical properties, High-Entropy photocatalysts represent an extremely promising area of research with substantial potential for environmental applications, including water splitting and carbon dioxide conversion.

Initial studies by Edalati et al. [21] have introduced a high-entropy oxide (HEO) with the following composition: $(TiZrHfNbTa)O_{11}$, that demonstrated noteworthy hydrogen evolution rates during photocatalytic water splitting. Such HEO system exhibits a superior absorption profile in the visible light region and a more favorable bandgap relative to its constituent cation oxides. Subsequent work by Akrami et al. [22] extended the application of $(TiZrHfNbTa)O_{11}$ to photocatalytic CO_2 conversion, with yields significantly exceeding those from traditional photocatalysts like anatase TiO_2 and $BiVO_4$ [23].

To improve absorption efficiency further, a novel high-entropy oxynitride was developed by Akrami et al. [24], showing a considerably reduced bandgap and enhanced photocatalytic stability. Additionally, two fluorite-structured HEOs synthesized by Anandkumar et al. [25] for water pollution treatment achieved significant photoreduction of hazardous Cr(VI) ions to Cr(III) ions and degradation of methylene blue dye.

Notably, rare earth-based HEOs seem to exhibit, other than unique optical properties, enhanced photocatalytic activities for the degradation of several pollutants, indicating their utility in both environmental remediation and chemical conversion.

References

[1] Callister Jr, W.D. & Rethwisch, D.G. (2020). Fundamentals of Materials Science and Engineering: An Integrated Approach. John Wiley & Sons.

[2] Ameta, R., Solanki, M.S., Benjamin, S. & Ameta, S.C. (2018). Photocatalysis. *In:* Advanced Oxidation Processes for Waste Water Treatment (pp. 135–175). Academic Press.

[3] Zhu, Z., Zhou, F. & Zhan, S. (2020). Enhanced antifouling property of fluorocarbon resin coating (PEVE) by the modification of g-C_3N_4/Ag_2WO_4 composite step-scheme photocatalyst. Applied Surface Science, 506, 144934.

[4] Duan, Z., Zhu, Y., Ren, P., Jia, J., Yang, S., Zhao, G., ... & Zhang, J. (2018). Non-UV activated superhydrophilicity of patterned Fe-doped TiO_2 film for anti-fogging and photocatalysis. Applied Surface Science, 452, 165–173.

[5] Li, J., Liu, Y., Zhu, Z., Zhang, G., Zou, T., Zou, Z., ... & Xie, C. (2013). A full-sunlight-driven photocatalyst with super long-persistent energy storage ability. Scientific Reports, 3(1), 2409.

[6] Rajagopal, R. & Ryu, K.S. (2018). Synthesis of rGO-doped Nb_4O_5–TiO_2 nanorods for photocatalytic and electrochemical energy storage applications. Applied Catalysis B: Environmental, 236, 125–139.

[7] Mokhtar Mohamed, M., Mousa, M.A., Khairy, M. & Amer, A.A. (2018). Nitrogen graphene: A new and exciting generation of visible light driven photocatalyst and energy storage application. ACS Omega, 3(2), 1801–1814.

[8] Herrmann, J.M. & Lacroix, M. (2010). Environmental photocatalysis in action for green chemistry. Kinetics and Catalysis, 51, 793–800.

[9] Gong, M., Xiao, S., Yu, X., Dong, C., Ji, J., Zhang, D. & Xing, M. (2019). Research progress of photocatalytic sterilization over semiconductors. RSC Advances, 9(34), 19278–19284.

[10] Ren, H., Koshy, P., Chen, W.F., Qi, S. & Sorrell, C.C. (2017). Photocatalytic materials and technologies for air purification. Journal of Hazardous Materials, 325, 340–366.

[11] Mamaghani, A.H., Haghighat, F. & Lee, C.S. (2020). Role of titanium dioxide (TiO_2) structural design/morphology in photocatalytic air purification. Applied Catalysis B: Environmental, 269, 118735.

[12] Al-Mamun, M.R., Kader, S., Islam, M.S. & Khan, M.Z.H. (2019). Photocatalytic activity improvement and application of $UV-TiO_2$ photocatalysis in textile wastewater treatment: A review. Journal of Environmental Chemical Engineering, 7(5), 103248.

[13] Ren, G., Han, H., Wang, Y., Liu, S., Zhao, J., Meng, X. & Li, Z. (2021). Recent advances of photocatalytic application in water treatment: A review. Nanomaterials, 11(7), 1804.

[14] Gusain, R., Kumar, N. & Ray, S.S. (2020). Factors influencing the photocatalytic activity of photocatalysts in wastewater treatment. Photocatalysts in Advanced Oxidation Processes for Wastewater Treatment, 229–270.

[15] Chen, T.K. & Wong, M.S. (2007). Structure and properties of reactively-sputtered AlxCoCrCuFeNi oxide films. Thin Solid Films, 516(2–4), 141–146.

[16] Yang, Y., Wang, W., Gan, G.Y., Shi, X.F. & Tang, B.Y. (2018). Structural, mechanical and electronic properties of (TaNbHfTiZr)C high entropy carbide under pressure: Ab initio investigation. Physica B: Condensed Matter, 550, 163–170.

[17] Sarkar, A., Eggert, B., Velasco, L., Mu, X., Lill, J., Ollefs, K., ... & Hahn, H. (2020). Role of intermediate 4f states in tuning the band structure of high entropy oxides. APL Materials, 8(5).

[18] Liu, Y., Gui, Z. & Liu, J. (2022). Research progress of light wavelength conversion materials and their applications in functional agricultural films. Polymers, 14(5), 851.

[19] Hanioka, M., Furukawa, Y. & Samata, H. (2019). Wavelength conversion characteristics of Y1–x–yErxYbyBaZn$_3$AlO$_7$. Optik, 180, 1043–1048.

[20] Zhang, J., Zhang, X., Li, Y., Du, Q., Liu, X. & Qi, X. (2019). High-entropy oxides $10La_2O_3$-$20TiO_2$-$10Nb_2O_5$-$20WO_3$-$20ZrO_2$ amorphous spheres prepared by containerless solidification. Materials Letters, 244, 167–170.

[21] Edalati, P., Shen, X.-F., Watanabe, M., Ishihara, T., Arita, M., Fuji, M. & Edalati, K. (2021). High-entropy oxynitride as a low-bandgap and stable photocatalyst for hydrogen production. Journal of Materials Chemistry A, 9(26), 15076–15086.

[22] Akrami, S., Murakami, Y., Watanabe, M., Ishihara, T., Arita, M., Fuji, M. & Edalati, K. (2022). Defective high-entropy oxide photocatalyst with high activity for CO_2 conversion. Applied Catalysis B: Environment and Energy, 303.

[23] Pan, Y., Liu, J.X., Tu, T.Z., Wang, W. & Zhang, G.J. (2023). High-entropy oxides for catalysis: A diamond in the rough. Chemical Engineering Journal, 451, 138659.

[24] Akrami, S., Edalati, P., Shundo, Y., Watanabe, M., Ishihara, T., Fuji, M. & Edalati, K. (2022). Significant CO_2 photoreduction on a high-entropy oxynitride. Chemical Engineering Journal, 449, 137800.

[25] Anandkumar, M., Lathe, A., Palve, A.M. & Deshpande, A.S. (2021). Single-phase $Gd_{0.2}La_{0.2}Ce_{0.2}Hf_{0.2}Zr_{0.2}O_2$ and $Gd_{0.2}La_{0.2}Y_{0.2}Hf_{0.2}Zr_{0.2}O_2$ nanoparticles as efficient photocatalysts for the reduction of Cr(vi) and degradation of methylene blue dye. Journal of Alloys and Compounds, 850.

Conclusions and Future Scenarios

In the vibrant field of materials science, the intriguing concept of high-entropy ceramics has recently emerged (in 2015), expanding upon the foundations laid by high-entropy alloys about a decade before. Such a novel class of ceramics is characterized by their rich and complex compositions, typically involving five or more constituting elements in almost equal proportions. These elements, primarily metal cations, are intricately woven into the structure of ceramic compounds such as variously-structured oxides, carbides, nitrides, borides and silicides. The hallmark of these novel materials is the entropic *chaos* resulting from the disordered arrangement of different cations within the same crystal lattice.

The properties of high-entropy ceramics are as diverse as their compositions. They are renowned for their remarkable thermal properties, but they also exhibit potential for unique electrical and magnetic properties, depending on the specific composition mix and crystal structure. This opens doors to a wide array of applications, ranging from protective coatings and high-temperature components in gas turbines to advanced roles in electronic and magnetic devices.

However, the path to harnessing the full potential of high-entropy ceramics is just at its beginning and it is not without its challenges: one of the primary issues to address lies in predicting and controlling their properties, given the intricate interplay among the diversity of constituent elements. However, this complexity makes the field of high-entropy ceramics a vibrant area of ongoing research, with many scientists continually exploring new compositions and techniques to tailor these materials for specific uses. Particularly, the research on high-entropy ceramics, being at the cutting edge of materials science, is expected to venture into several promising directions in the near future:

- **Compositional design** – A primary area of research exploration involves expanding the range of compositions used in high-entropy ceramics. This includes experimenting with different combinations of metal cations and exploring the inclusion of non-metal elements different from the oxygen. Advanced computational methods, like machine learning and predictive modeling, could play a significant role in identifying novel, promising compositions and in understanding the relationship between composition, structure, and properties in high-entropy ceramics.

- **Properties optimization** – Tailoring and optimizing specific properties of high-entropy ceramics for particular applications is another ongoing research key focus. For instance, enhancing and tuning their promising thermal, electrical, optical and magnetic properties could open up new applications for high-entropy ceramics as thermal barrier coatings, batteries' electrodes, insulators, catalysts and so on. Similarly, improving mechanical properties, especially for high-entropy carbides, borides and nitrides, could make these ceramics suitable for highly specific technological applications.
- **Biocompatibility** – Studies on the biocompatibility of high-entropy ceramics might open up applications in the biomedical field too, as well as ascertain their intrinsic safety for humans and for the ecosystem.
- **Understanding and modeling** – Understanding how different elemental compositions influence the final entropy-stabilized structure (especially for complex systems like perovskites) and, in turn, the material properties, will be key to custom-designing high-entropy ceramics for real-life technological applications.
- **Scalability and manufacturing** – Finally, future research on high-entropy oxides should also focus on developing cost-effective and scalable manufacturing processes for high-entropy ceramics. This includes optimizing synthesis and sintering techniques, as well as exploring novel fabrication methods like additive manufacturing (3D printing) for ceramics.

Definitely, as research in the field of high-entropy ceramics advances, the author expects that such a noel and intriguing class of materials will revolutionize a wide range of industrial and technological applications, maybe even unlocking the full potential of high-temperature superconductors. Thus, the exploration of high-entropy ceramics is part of a larger ongoing shift in materials science, where the principles of the 'high entropy approach' is leading to the discovery of materials with unprecedented properties, significantly enriching our understanding of nature (the most able *high-entropy architect*) and opening up new possibilities for the humankind technological advancement.

Index